Kate Gathercole

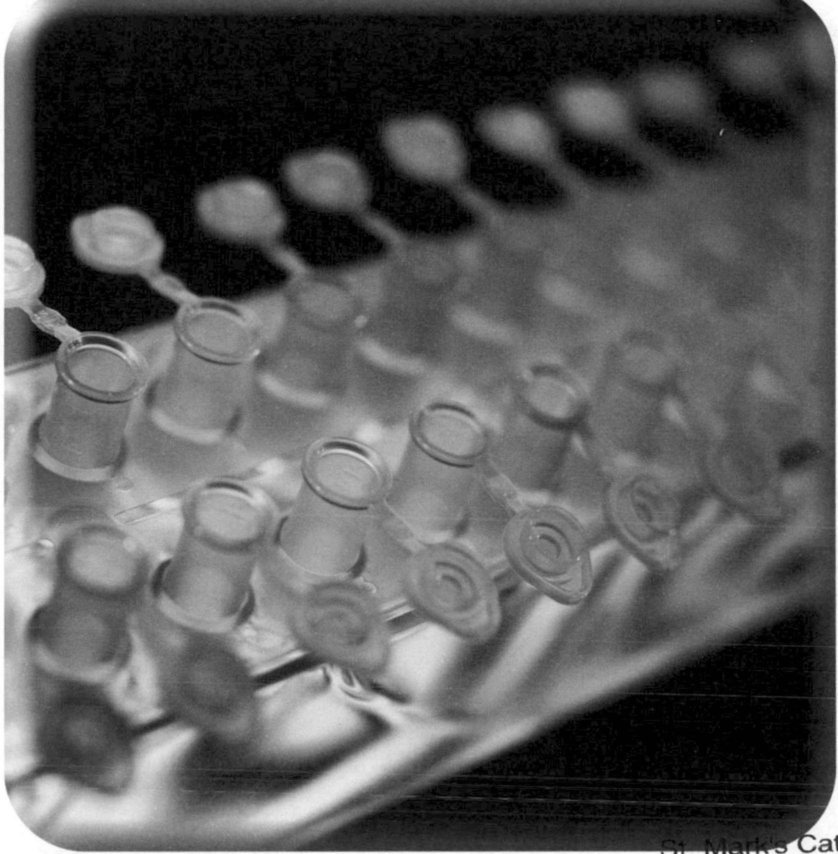

REVISION PLUS

AQA
GCSE Chemistry
Revision and Classroom Companion

Contents

N.B. The numbers in brackets correspond to the reference numbers in the AQA Chemistry specification.

The new AQA GCSE Chemistry specification incorporates two types of content:

- **Science Content** (example shown opposite). This includes all the scientific explanations and evidence that you need to be able to recall in your exams (objective tests or written exams). It is covered on pages 11–98 of the revision guide.
- **How Science Works** (example shown opposite). This is a set of key concepts, relevant to all areas of science. It is concerned with how scientific evidence is obtained and the effect it has on society. More specifically, it covers:
  - the relationship between scientific evidence and scientific explanations and theories
  - the practices and procedures used to collect scientific evidence
  - the reliability and validity of scientific evidence
  - the role of science in society and the impact it has on our lives
  - how decisions are made about the use of science and technology in different situations, and the factors affecting these decisions

Because they are interlinked, your teacher may have taught the two types of content together in your science lessons. Likewise, the questions on your exam papers are likely to combine elements from both types of content. To answer them you will need to recall the relevant scientific facts and draw on your knowledge of how science works.

The key concepts from How Science Works are summarised in this section of the revision guide. You should be familiar with all of them, especially the practices and procedures used to collect scientific data (from all your practical investigations). Make sure you work through them all. Make a note of anything you are unsure about and then ask your teacher for clarification.

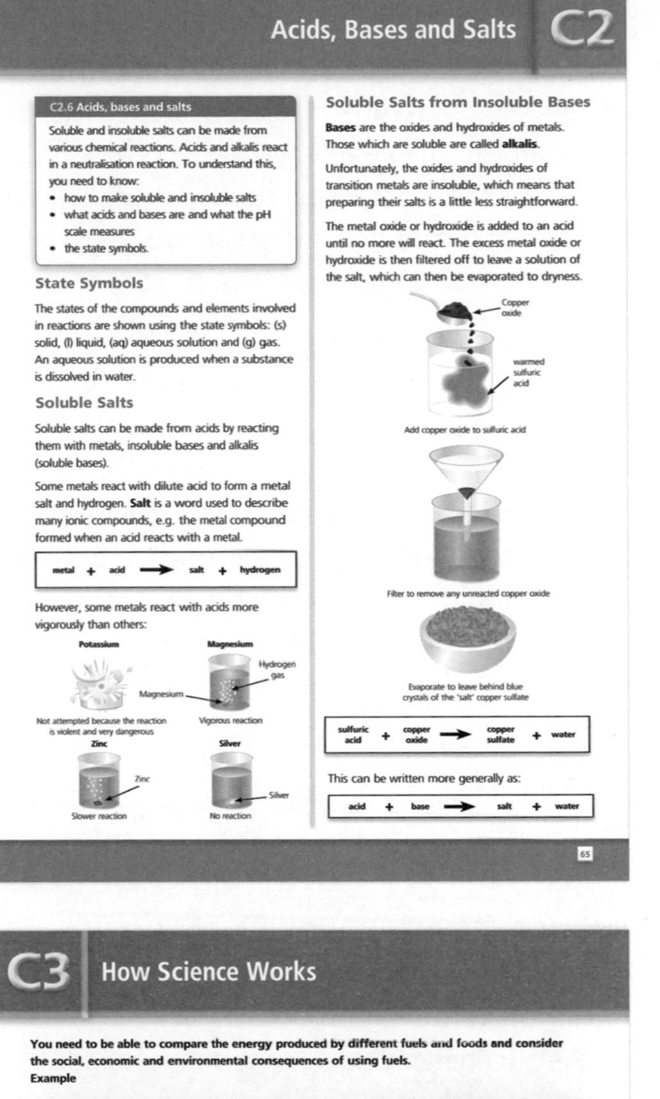

Acids, Bases and Salts C2

C2.6 Acids, bases and salts

Soluble and insoluble salts can be made from various chemical reactions. Acids and alkalis react in a neutralisation reaction. To understand this, you need to know:
- how to make soluble and insoluble salts
- what acids and bases are and what the pH scale measures
- the state symbols.

State Symbols

The states of the compounds and elements involved in reactions are shown using the state symbols: (s) solid, (l) liquid, (aq) aqueous solution and (g) gas. An aqueous solution is produced when a substance is dissolved in water.

Soluble Salts

Soluble salts can be made from acids by reacting them with metals, insoluble bases and alkalis (soluble bases).

Some metals react with dilute acid to form a metal salt and hydrogen. **Salt** is a word used to describe many ionic compounds, e.g. the metal compound formed when an acid reacts with a metal.

metal + acid → salt + hydrogen

However, some metals react with acids more vigorously than others:

Soluble Salts from Insoluble Bases

Bases are the oxides and hydroxides of metals. Those which are soluble are called **alkalis**.

Unfortunately, the oxides and hydroxides of transition metals are insoluble, which means that preparing their salts is a little less straightforward.

The metal oxide or hydroxide is added to an acid until no more will react. The excess metal oxide or hydroxide is then filtered off to leave a solution of the salt, which can then be evaporated to dryness.

sulfuric acid + copper oxide → copper sulfate + water

This can be written more generally as:

acid + base → salt + water

65

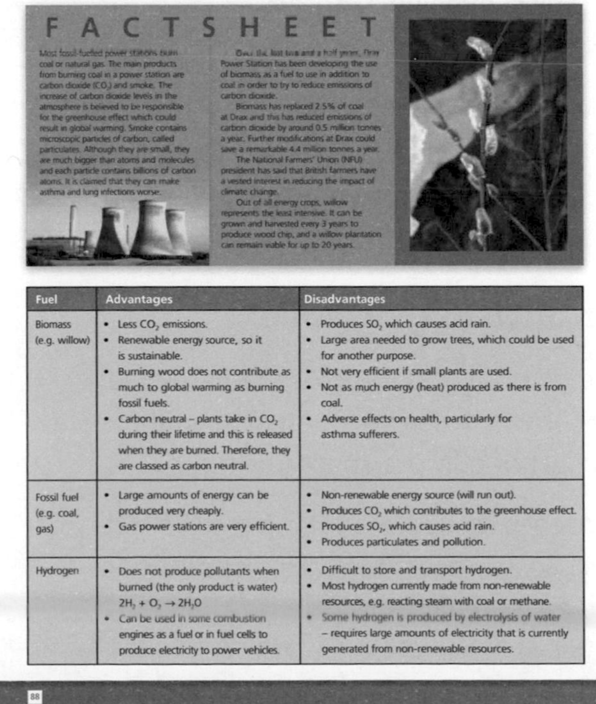

C3 How Science Works

You need to be able to compare the energy produced by different fuels and foods and consider the social, economic and environmental consequences of using fuels.
Example

FACTSHEET

Most fossil-fuelled power stations burn coal or natural gas. The main products from burning coal in a power station are carbon dioxide (CO_2) and smoke. The increase of carbon dioxide levels in the atmosphere is believed to be responsible for the greenhouse effect which could result in global warming. Smoke contains microscopic particles of carbon, called particulates. Although they are small, they are much bigger than atoms and molecules and each particle contains billions of carbon atoms. It is claimed that they can make asthma and lung infections worse.

Over the last two and a half years, Drax Power Station has been developing the use of biomass as a fuel to use in addition to coal in order to try to reduce emissions of carbon dioxide.

Biomass has replaced 2.5% of coal at Drax and this has reduced emissions of carbon dioxide by around 0.5 million tonnes a year. Further modifications at Drax could save a remarkable 4.4 million tonnes a year.

The National Farmers' Union (NFU) president has said that British farmers have a vested interest in reducing the impact of climate change.

Out of all energy crops, willow represents the least intensive. It can be grown and harvested every 3 years to produce wood chip, and a willow plantation can remain viable for up to 20 years.

Fuel	Advantages	Disadvantages
Biomass (e.g. willow)	• Less CO_2 emissions. • Renewable energy source, so it is sustainable. • Burning wood does not contribute as much to global warming as burning fossil fuels. • Carbon neutral – plants take in CO_2 during their lifetime and this is released when they are burned. Therefore, they are classed as carbon neutral.	• Produces SO_2 which causes acid rain. • Large area needed to grow trees, which could be used for another purpose. • Not very efficient if small plants are used. • Not as much energy (heat) produced as there is from coal. • Adverse effects on health, particularly for asthma sufferers.
Fossil fuel (e.g. coal, gas)	• Large amounts of energy can be produced very cheaply. • Gas power stations are very efficient.	• Non-renewable energy source (will run out). • Produces CO_2 which contributes to the greenhouse effect. • Produces SO_2, which causes acid rain. • Produces particulates and pollution.
Hydrogen	• Does not produce pollutants when burned (the only product is water) $2H_2 + O_2 \rightarrow 2H_2O$ • Can be used in some combustion engines as a fuel or in fuel cells to produce electricity to power vehicles.	• Difficult to store and transport hydrogen. • Most hydrogen currently made from non-renewable resources, e.g. reacting steam with coal or methane. • Some hydrogen is produced by electrolysis of water – requires large amounts of electricity that is currently generated from non-renewable resources.

88

How Science Works Overview

How Science Works

The AQA GCSE Chemistry specification includes activities for each sub-section of science content. These require you to apply your knowledge of how science works and will help develop your skills when it comes to evaluating information, developing arguments and drawing conclusions.

These activities are dealt with on the How Science Works pages throughout the revision guide. Make sure you work through them all, because questions relating to the skills, ideas and issues covered on these pages could easily come up in the exam. Bear in mind that these pages are designed to provide a starting point from which you can begin to develop your own ideas and conclusions. They are not meant to be definitive or prescriptive.

Practical tips on how to evaluate information are included in this section, on page 9.

What is the Purpose of Science?

Science attempts to explain the world we live in. The role of a scientist is to collect evidence through investigations to:

- explain phenomena (i.e. explain how and why something happens)
- solve problems.

Scientific knowledge and understanding can lead to the development of new technologies (e.g. in medicine and industry), which have a huge impact on society and the environment.

Scientific Evidence

The purpose of evidence is to provide facts that answer a specific question, and therefore support or disprove an idea or theory.

In science, evidence is often based on data that has been collected by making observations and measurements.

A scientifically literate citizen should be equipped to question the evidence used in decision making. Evidence must therefore be approached with a critical eye. It is necessary to look closely at:

- how measurements have been made
- whether opinions drawn are based on valid and reproducible evidence rather than non-scientific ideas (prejudices and hearsay)
- whether the evidence is reproducible, i.e. it can be reproduced by others
- whether the evidence is valid, i.e. is reproducible, repeatable and answers the original question.

N.B. If data is not reproducible or repeatable, it cannot be valid.

To ensure scientific evidence is repeatable, reproducible and valid, scientists employ a range of ideas and practices that relate to the following:

1. **Observations** – how we observe the world.
2. **Investigations** – designing investigations so that patterns and relationships can be identified.
3. **Measurements** – making measurements by selecting and using instruments effectively.
4. **Presenting data** – presenting and representing data.
5. **Conclusions** – identifying patterns and relationships and making suitable conclusions.
6. **Evaluation** – considering the validity of data and appropriateness of methods used.

These six key ideas are covered in more detail on the following pages.

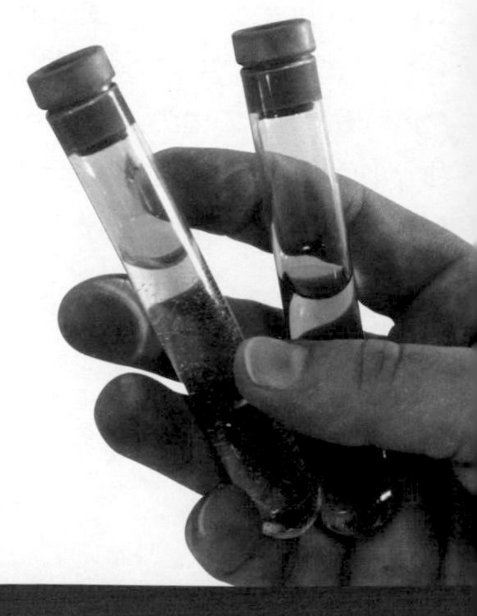

How Science Works Overview

❶ Observations

Most scientific investigations begin with an observation, i.e. a scientist observes an event or phenomenon and decides to find out more about how and why it happens.

The first step is to develop a **hypothesis**, i.e. to suggest an explanation for the phenomenon. Hypotheses normally propose a relationship between two or more **variables** (factors that can change). Hypotheses are based on careful observations and existing scientific knowledge, and often include a bit of creative thinking.

The hypothesis is used to make a prediction, which can be tested through scientific investigation. The data collected during the investigation might support the hypothesis, show it to be untrue, or lead to the development of a new hypothesis.

Example
A biologist **observes** that freshwater shrimp are only found in certain parts of a stream.

The biologist uses current scientific knowledge of freshwater shrimp behaviour and water flow to develop a **hypothesis**, which relates the distribution of shrimp (first variable) to the rate of water flow (second variable).

Based on this hypothesis, the biologist **predicts** that shrimp can only be found in areas of the stream where the flow rate is beneath a certain value.

The prediction is **investigated** through a survey that looks for the presence of shrimp in different parts of the stream, which represent a range of different flow rates.

The **data** shows that shrimp are only present in parts of the stream where the flow rate is below a certain value (i.e. the data supports the hypothesis). However, it also shows that shrimp are not *always* present in parts of the stream where the flow rate is below this value.

As a result, the biologist realises that there must be another factor affecting the distribution of shrimp. So, he **refines his hypothesis** to relate the distribution of shrimp (first variable) to the concentration of oxygen in the water (second variable) in parts of the stream where there is a slow flow rate.

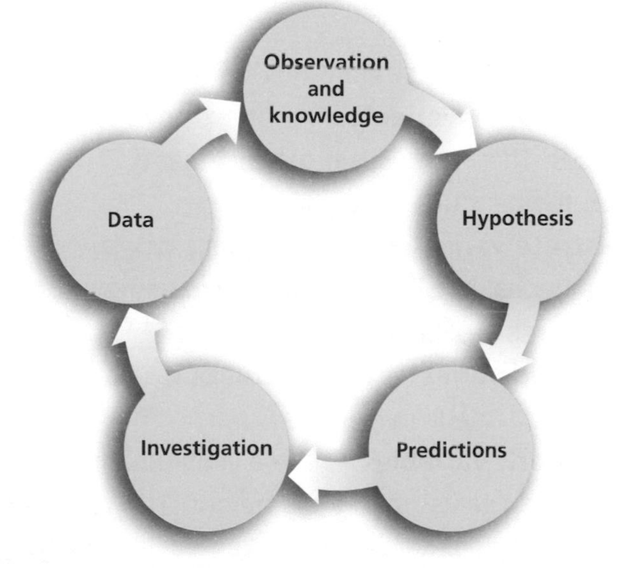

If new observations or data do not match existing explanations or theories (e.g. if unexpected behaviour is displayed) they need to be checked for reliability and validity.

In some cases it turns out that the new observations and data are valid, so existing theories and explanations have to be revised or amended. This is how scientific knowledge gradually grows and develops.

2 Investigations

An investigation involves collecting data to try to determine if there is a relationship between two variables. A variable is any factor that can take different values (i.e. any factor that can change). In an investigation there are three types of variables:

- **Independent variable**, which is changed or selected by the person carrying out the investigation. In the shrimp example on page 5, the independent variable is the flow rate of the water.
- **Dependent variable**, which is measured each time a change is made to the independent variable, to see if it also changes. In the shrimp example on page 5, the dependent variable is the number of shrimp.
- **Outside variables**, which are variables that could affect the dependent variable and must be controlled.

For a measurement to be valid it must measure only the appropriate variable.

Variables can have different types of values:

- **Continuous variables** – can take any numerical values. These are usually measurements, e.g. temperature or height.
- **Discrete variables** – can only take whole-number values. These are usually quantities, e.g. the number of shrimp in a population.
- **Ordered variables** – have relative values, e.g. small, medium or large.
- **Categoric variables** – have a limited number of specific values, e.g. the different breeds of dog: dalmatian, cocker spaniel, labrador, etc.

Numerical values tend to be more powerful and informative than ordered variables and categoric variables.

An investigation tries to establish if an observed link between two variables is one of the following:

- **Causal** – a change in one variable causes a change in the other, e.g. in a chemical reaction the rate of reaction (dependent variable) increases when the temperature of the reactants (independent variable) is increased.
- **Due to association** – the changes in the two variables are linked by a third variable. For example, a link between the change in pH of a stream (first variable) and a change in the number of different species found in the stream (second variable), may be the effect of a change in the concentration of atmospheric pollutants (third variable).
- **Due to 'chance occurrence', i.e. coincidence** – the change in the two variables is unrelated; it is coincidental. For example, in the 1940s the number of deaths due to lung cancer increased and the amount of tar being used in road construction increased. However, one increase *did not* cause the other increase.

2 Investigations (cont)

Fair Tests

A fair test is one in which the only factor that can affect the dependent variable is the independent variable. Any other variables (outside variables) that could influence the results are kept the same.

A fair test is much easier to achieve in the laboratory than in the field, where conditions (e.g. weather) cannot always be physically controlled. The impact of outside variables, like the weather, has to be reduced by ensuring that all measurements are affected by the variable in the same way. For example, if you were investigating the effect of different fertilisers on the growth of tomato plants, all the plants would need to be grown in a place where they were subject to the same weather conditions.

If a survey is used to collect data, the impact of outside variables can be reduced by ensuring that the individuals in the sample are closely matched. For example, if you were investigating the effect of smoking on life expectancy, the individuals in the sample would all need to have a similar diet and lifestyle to ensure that those variables did not affect the results.

Control groups are often used in biological research. For example, in some drugs trials a placebo (a dummy pill containing no medicine) is taken by one group of volunteers (the control group) and the drug is taken by another group. By comparing the two groups, scientists can establish if the drug (the independent variable) is the only variable affecting the volunteers and therefore whether or not it is a fair test.

Selecting Values of Variables

Care is needed in selecting the values of variables to be recorded in an investigation. For example, if you were investigating the effect of fertilisers on plant growth, you would need a range of fertiliser concentration that would give a measurable range of growth. Too narrow a range of concentration may fail to give any noticeable difference in growth. A trial run often helps to identify appropriate values to be recorded, such as the number of repeated readings needed and their range and interval.

Repeatability, Reproducibility and Validity

Repeatability measures how consistent results are in a single experiment. For example, a student measures the time taken for magnesium to react with hydrochloric acid. She repeats the test five times and finds that for each test the reaction takes 45 seconds. We can say that repeatability is high and the results are accurate.

In practice, it is unlikely that all five tests will give exactly the same result and therefore the mean (average) of a set of repeated measurements is often calculated to overcome small variations and get a best estimate of a true result.

An accurate measurement is one that is close to the true value. The purpose of an investigation will determine how accurate the data collected needs to be. For example, measurements of blood alcohol levels must be accurate enough to determine if a person is legally fit to drive.

Reproducibility measures the ability of an experiment to produce results that are the same each time it is carried out. For example, a whole class of students carry out the experiment above with magnesium and hydrochloric acid. The results can be said to be reproducible, and therefore reliable, if all students' results are very close to each other.

Validity questions if the results obtained can be used to prove or disprove the original hypothesis. It must consider the design of the experiment, the extent to which variables have been controlled and the reliability of results. The data collected must be precise enough (i.e. to an appropriate number of decimal places) to form a valid conclusion.

3 Measurements

When making measurements, errors may occur that affect the repeatability, reproducibility and validity of the results. These may be due to:

- **Variables that have not been controlled** – these may be variables that are beyond control.
- **Human error** – when making measurements, random errors can occur due to a lapse in concentration. Random errors can also result from inconsistent application of a technique. Systematic (repeated) errors can result from an instrument not being calibrated correctly or repeatedly being misused, or from consistent misapplication of a technique.
- **The resolution of the instruments used** – the resolution of an instrument is determined by the smallest change in value it can detect. For example, the resolution of bathroom scales is insufficient to detect the changes in mass of a small baby, whereas the scales used by a midwife have a higher resolution. It is therefore important to select instruments with an appropriate resolution for the task.

There will always be some variation in the actual value of a variable no matter how hard we try to repeat an event. For example, if the same model parachute was dropped twice from the same height, in the same laboratory and how long it took to fall to the ground was timed, it is very unlikely both drops would take exactly the same time. However, the two results should be close to each other. Any anomalous values should be examined to try to identify the cause. If the result is a product of a poor measurement, it should be ignored.

4 Presenting Data

When presenting data, two terms are frequently used:

- the **mean** (or average) – this is the sum of all the measurements divided by the number of measurements taken
- the **range** – this refers to the maximum and minimum values and the difference between them.

To explain the relationship between two or more variables, data can be presented in such a way that the pattern is more evident. The type of presentation used will depend on the type of variable represented.

Tables are an effective way of displaying data, but are limited in how much they can tell us about the design of an investigation.

Height (cm)	127	165	149	147	155	161	154	138	145
Shoe size	5	8	5	6	5	5	6	4	5

Bar charts are used to display data when one of the variables is categoric. They can also be used when one of the variables is discrete.

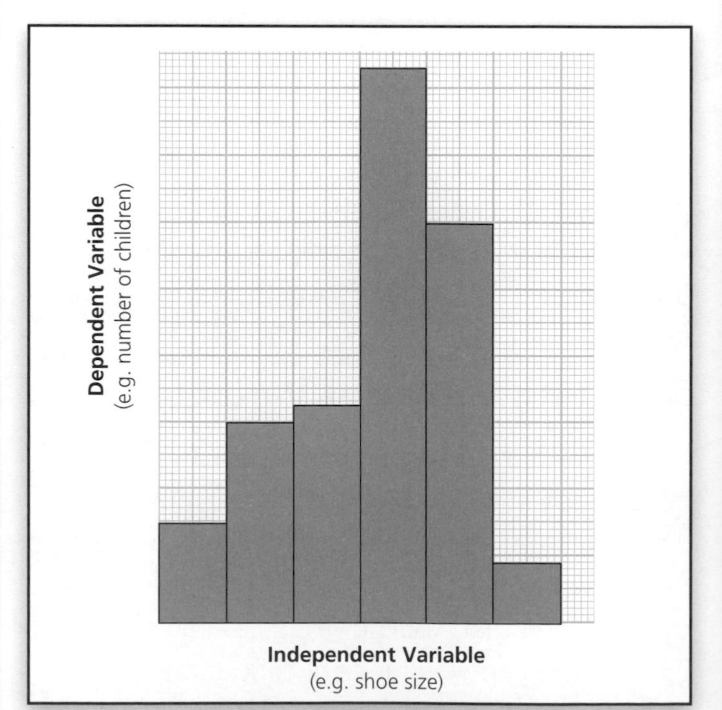

Scattergrams can be used to show an association between two variables. This can be made clearer by including a line of best fit.

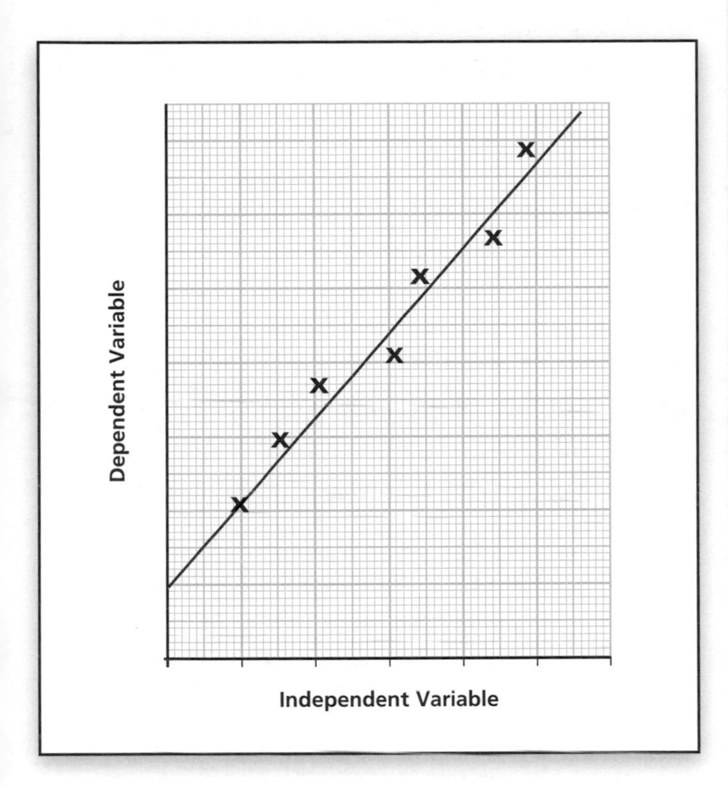

Line graphs are used when both the dependent and independent variables are continuous or discrete.

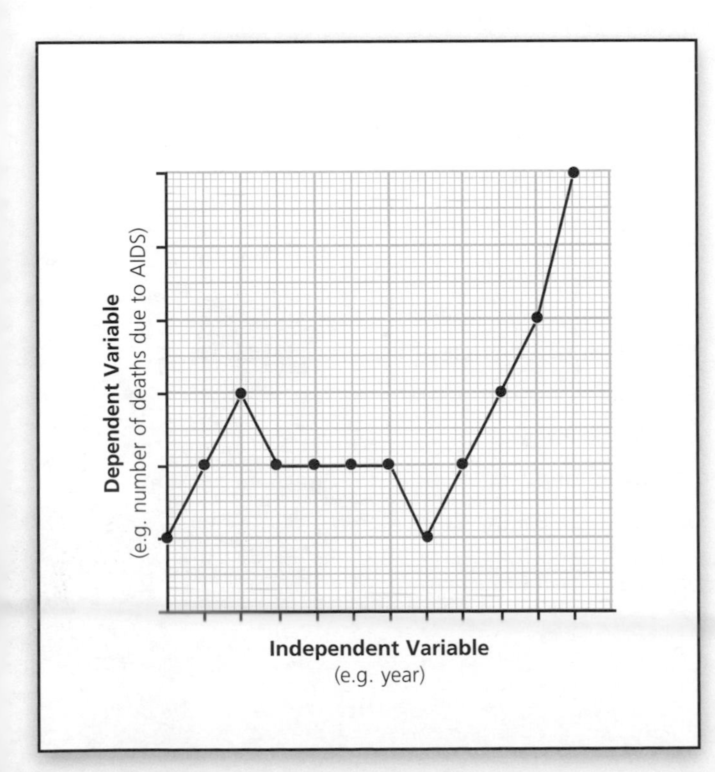

5 Conclusions

The patterns and relationships observed in the data represent what has happened in the investigation. However, it is necessary to look at the patterns and trends between the variables (bearing in mind limitations of the data), in order to draw conclusions.

Conclusions should:

- describe the patterns and relationships between variables
- take all the data into account
- make direct reference to the original hypothesis / prediction.

Conclusions should not:

- be influenced by anything other than the data collected
- disregard any data (other than anomalous values)
- include any speculation.

6 Evaluation

In evaluating a whole investigation, the repeatability, reproducibility and validity of the data obtained must be considered. This should take into account the original purpose of the investigation and the appropriateness of methods and techniques used in providing data to answer the original question.

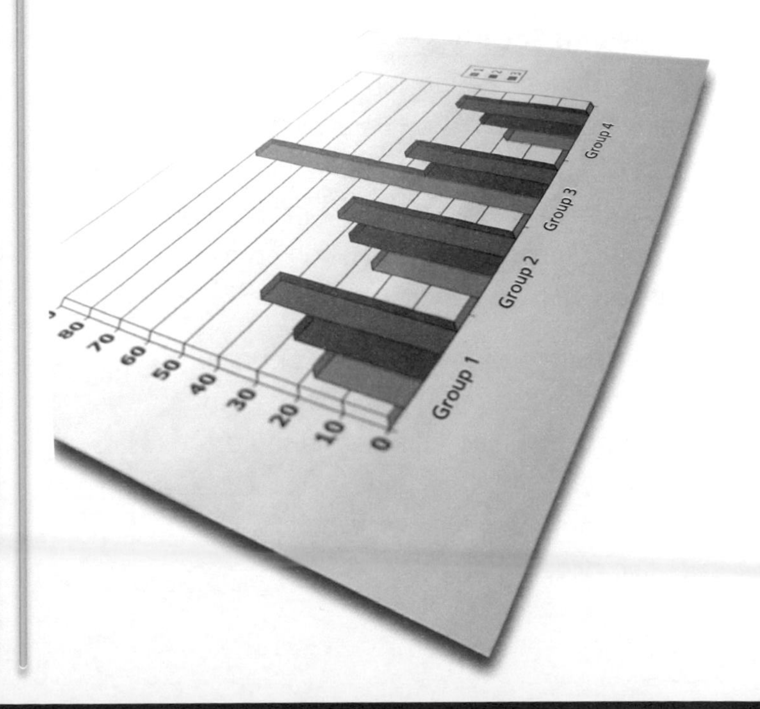

Societal Aspects of Scientific Evidence

Judgements and decisions relating to social-scientific issues may not be based on evidence alone. Sometimes other factors may have an influencing role. For example:

- **Bias** – evidence must be scrutinised for any potential bias on the part of the experimenter. Biased information might include incomplete evidence or may try to influence how you interpret the evidence. For example, this could happen if funding for the investigation came from a party with a vested interest, e.g. a drug company wanting to highlight the benefits of their new drug but downplay the side effects in order to increase sales.
- **Weight of evidence** – evidence can be given undue weight or be dismissed too lightly due to:
  - political significance – if the consequences of the evidence were likely to provoke public or political unrest or disquiet, the evidence may be downplayed
  - status of the experimenter – evidence is likely to be given more weight if it comes from someone with academic or professional status or who is considered to be an expert in that particular field.

Science and Society

Scientific understanding can lead to technological developments, which can be exploited by different groups of people for different reasons. For example, the successful development of a new drug benefits the drug company financially and improves the quality of life for patients.

The applications of scientific and technological developments can raise certain issues. An 'issue' is an important question that is in dispute and needs to be settled. Decisions made by individuals and society about these issues may not be based on scientific evidence alone.

Social issues are concerned with the impact on the human population of a community, city, country or even the world.

Economic issues are concerned with money and related factors like employment and the distribution of resources. There is often an overlap between social and economic issues.

Environmental issues are concerned with the impact on the planet, i.e. its natural ecosystems and resources.

Ethical issues are concerned with what is morally right and wrong, i.e. they require a value judgement to be made about what is acceptable. As society is underpinned by a common belief system, there are certain actions that can never be justified. However, because the views of individuals are influenced by lots of different factors (e.g. faith and personal experience) there are also lots of grey areas.

Limitations of Scientific Evidence

Science can help us in lots of ways but it cannot supply all the answers. We are still finding out about things and developing our scientific knowledge. There are some questions that we cannot answer, maybe because we do not have enough reproducible, repeatable and valid evidence.

There are some questions that science cannot answer at all. These tend to be questions relating to ethical issues, where beliefs and opinions are important, or to situations where we cannot collect reliable and valid scientific evidence. In other words, science can often tell us *whether* something can be done or not and *how* it can be done, but it cannot tell us if it *should* be done.

C1.1 The fundamental ideas in chemistry

Atoms are the building blocks of chemistry. Atoms contain protons, neutrons and electrons. Elements are substances containing only one type of atom – when elements react they produce compounds. Elements are grouped together in the Periodic Table. To understand this, you need to know:

- what atoms, elements and compounds are
- what the subatomic particles in atoms are and how electrons are arranged
- how elements are arranged in the Periodic Table (including knowledge of alkali metals, halogens and noble gases)
- how compounds are formed.

Atoms

All substances are made of **atoms** (very small particles). Each atom has a small central **nucleus** made up of **protons** and **neutrons** that is surrounded by **electrons**.

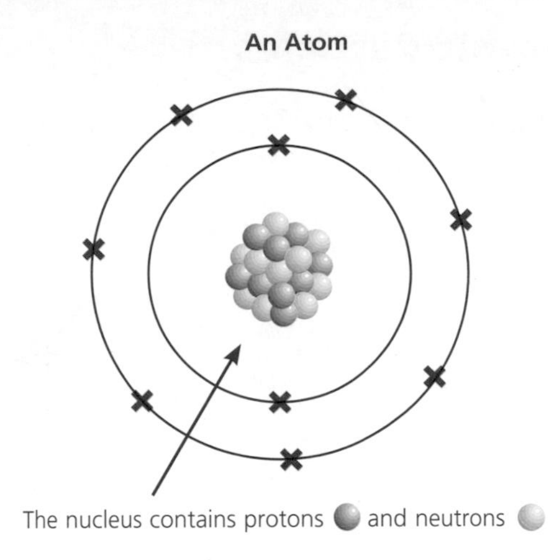

Elements

A substance that contains only one sort of atom is called an **element**. There are over 100 different elements.

The atoms of each element are represented by a different chemical symbol, for example, O for oxygen, Na for sodium, C for carbon, and Fe for iron. Elements are arranged in the Periodic Table (see below). The groups in the Periodic Table contain elements that have similar properties.

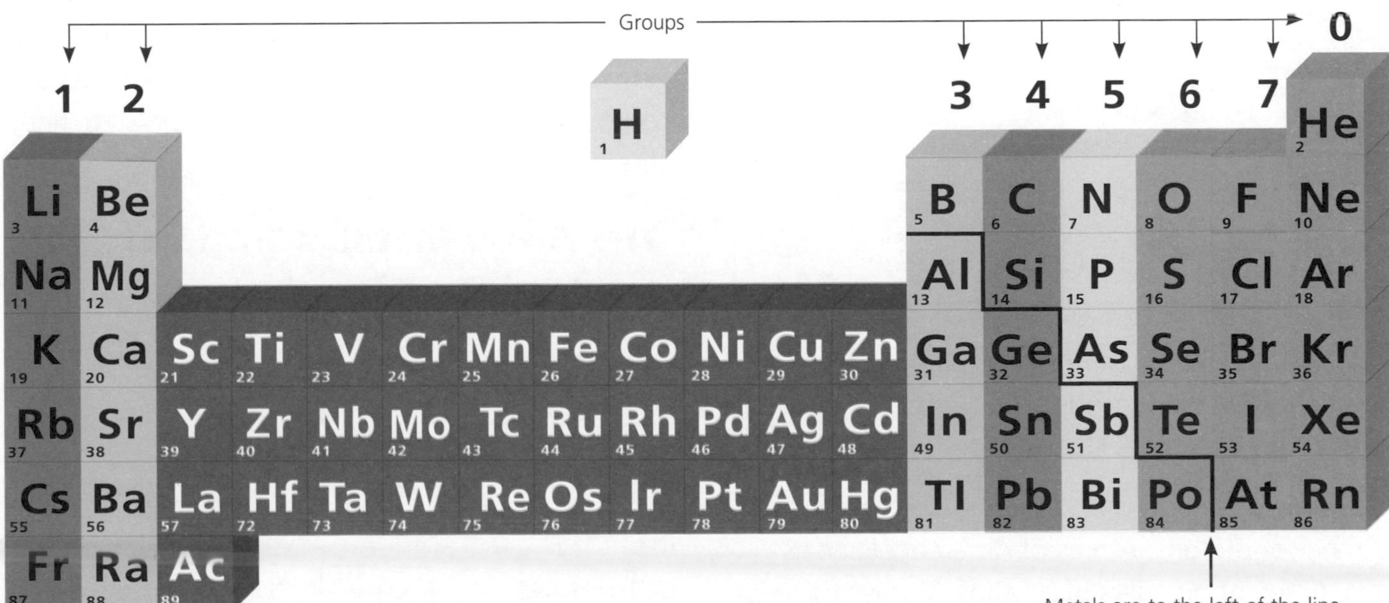

C1 The Fundamental Ideas in Chemistry

Subatomic Particles

The diagram below shows the subatomic particles in an atom of helium.

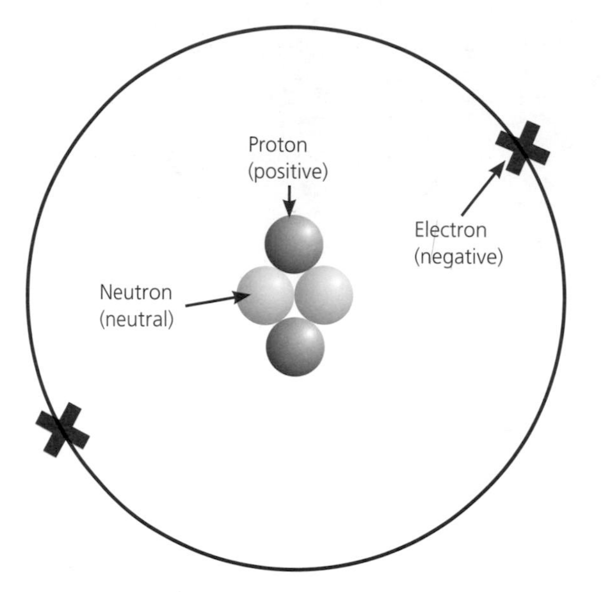

Atoms have a small central nucleus that is made up of protons and neutrons. The nucleus is surrounded by electrons. Protons, neutrons and electrons have relative electrical charges.

Atomic Particle	Relative Charge
Proton	+1
Neutron	0
Electron	–1

All atoms of a particular element have an equal number of protons and electrons, which means that atoms have no overall charge.

All atoms of a particular element have the same number of protons. Atoms of different elements have different numbers of protons. This is known as the **atomic number**.

Elements are arranged in the modern Periodic Table in order of increasing atomic number.

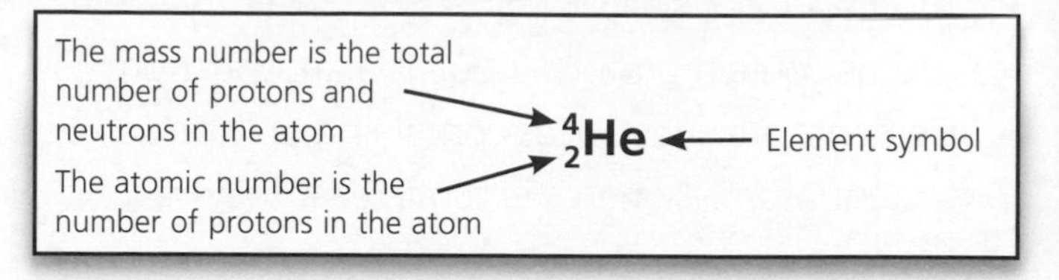

Electron configuration tells us how the electrons are arranged around the nucleus in energy levels or shells.

The electrons in an atom occupy the lowest available energy levels (i.e. the innermost available shells).

- The first level or shell can hold a maximum of 2 electrons.
- The energy levels or shells after this can hold a maximum of 8 electrons.
- The energy of the shells increases as they get further from the nucleus.

We write electron configurations as a series of numbers, e.g. oxygen is 2,6 and aluminium is 2,8,3.

The Periodic Table arranges the elements in terms of their electronic structure. Elements in the same group (column) have the same number of electrons in their outer shell (this number is the same as the group number). Elements in the same group therefore have similar properties. From left to right, across each period (row), a particular energy level is gradually filled with electrons. In the next period, the next energy level is filled, etc.

The Alkali Metals (Group 1)

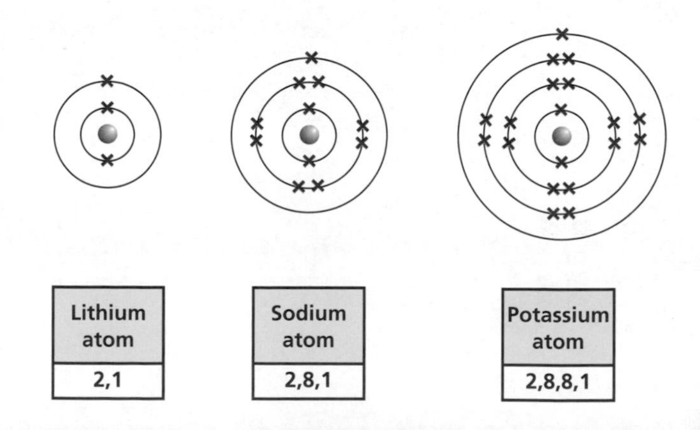

The alkali metals all have similar properties because their atoms have the same number of electrons (one) in their outermost shell, i.e. the highest occupied energy level contains one electron. They react with non-metal elements to form ionic compounds, in which the metal ions have a single positive charge.

Because the alkali metals all have one electron in their outer shell, they react in a similar way to each other. All the alkali metals react quite vigorously when you add them to water. They all produce an alkaline solution (a metal hydroxide) and hydrogen gas. For example:

sodium	+	water	→	sodium hydroxide	+	hydrogen
potassium	+	water	→	potassium hydroxide	+	hydrogen

The alkali metals will all react with oxygen to produce a metal oxide. For example:

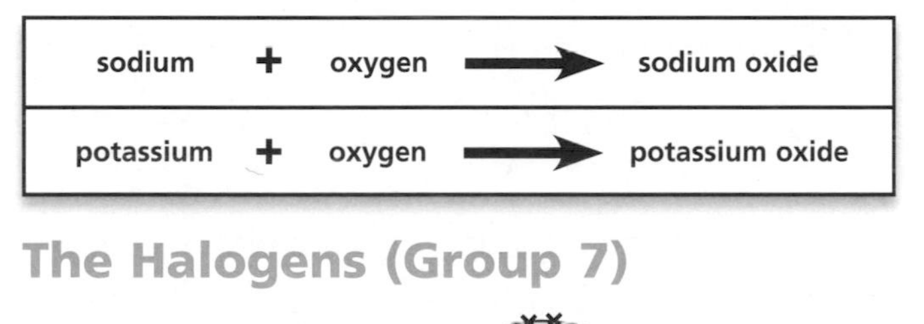

sodium	+	oxygen	→	sodium oxide
potassium	+	oxygen	→	potassium oxide

The Halogens (Group 7)

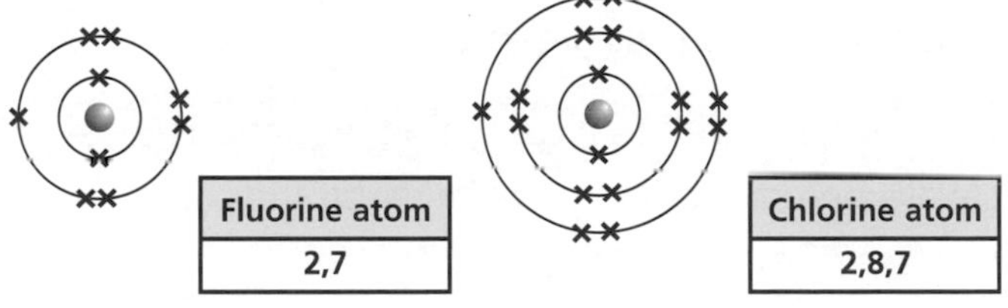

The halogens all have similar properties because their atoms have the same number of electrons (7) in their outermost shell, i.e. the highest occupied energy level contains 7 electrons.

They react with alkali metals to form ionic compounds, in which the halide ions have a single negative charge.

Noble Gases (Group 0)

The noble gases are inert (unreactive) gases. They are unreactive because they have a stable electron arrangement. The outer shell of each noble gas atom is full. Helium has 2 outer electrons. All other noble gases have 8 outer electrons.

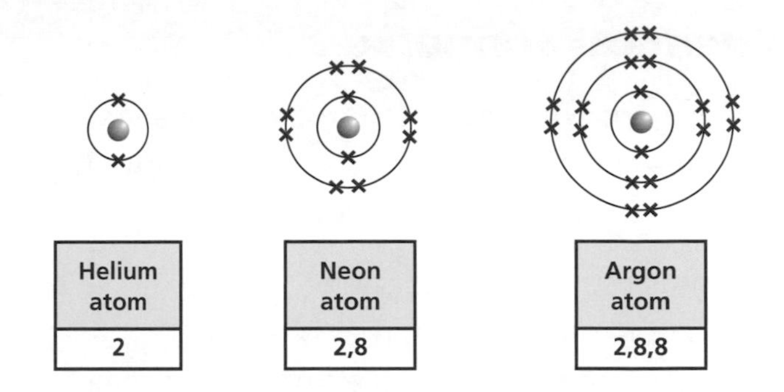

Mixtures and Compounds

A **mixture** consists of two or more elements or compounds that are not chemically combined. The properties of the substances remain unchanged and specific to each substance.

Compounds are substances in which the atoms of two or more elements are chemically combined (not just mixed together). The atoms are held together by chemical bonds.

Atoms can form chemical bonds by:

- sharing electrons (covalent bonds)
- gaining or losing electrons (ionic bonds).

Compounds formed from the reaction of metal atoms and non-metal atoms consist of ions. The metal atoms lose electrons to form positive ions. For example, a sodium atom (Na) loses an electron to become a sodium ion (Na^+). The non-metal atoms gain electrons to form negative ions. For example, a chlorine atom (Cl) gains an electron to form a chloride ion (Cl^-). These compounds are held together by ionic bonds, e.g. sodium chloride (NaCl).

Compounds formed from only non-metal atoms consist of molecules. In these, atoms share electrons and are held together by covalent bonds, e.g. carbon dioxide (CO_2). Bonding is dealt with more thoroughly in Unit C2.

Either way, when atoms form chemical bonds the arrangement of the outermost shell of electrons changes resulting in each atom getting a complete outer shell of electrons.

For most atoms this is eight electrons but for hydrogen it is only two.

Chemical Formulae

Compounds are represented by a combination of numbers and chemical symbols called a **formula**, e.g. ZnO or $2H_2SO_4$.

Chemists use formulae to show:

- the different elements in a compound
- the ratio of atoms of each element in the compound.

In chemical formulae, the position of the numbers tells you what is multiplied. Smaller numbers that sit below the line (subscripts) only multiply the symbol that comes immediately before it, and large numbers that are the same size as the letters multiply all the symbols that come after.

For example:

- H_2O means (2 × H) + (1 × O)
- 2NaOH means 2 × (NaOH) **or** (2 × Na) + (2 × O) + (2 × H)
- $Ca(OH)_2$ means (1 × Ca) + (2 × O) + (2 × H).

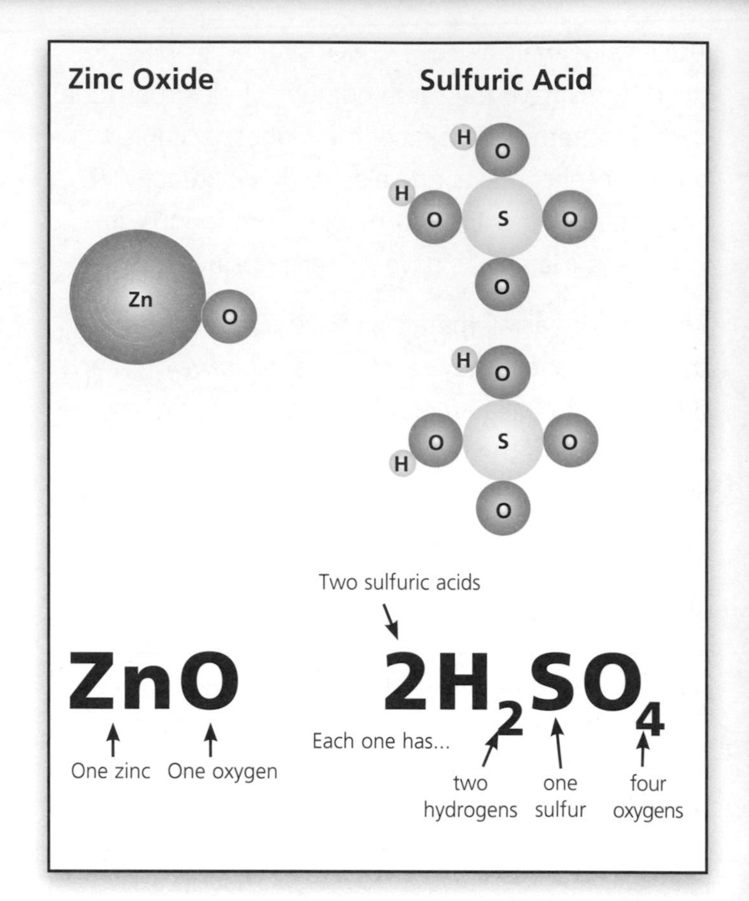

Chemical Reactions

You can show what has happened during a reaction by writing a **word equation** showing the substances that react (the **reactants**) on one side of the equation and the new substances formed (the **products**) on the other.

The total mass of the products of a chemical reaction is always equal to the total mass of the reactants.

This is because the products of a chemical reaction are made up from exactly the same atoms as the reactants – no atoms are lost or made!

That means that chemical symbol equations must always be balanced – there must be the same number of atoms of each element on the reactant side of the equation as there is on the product side.

Example

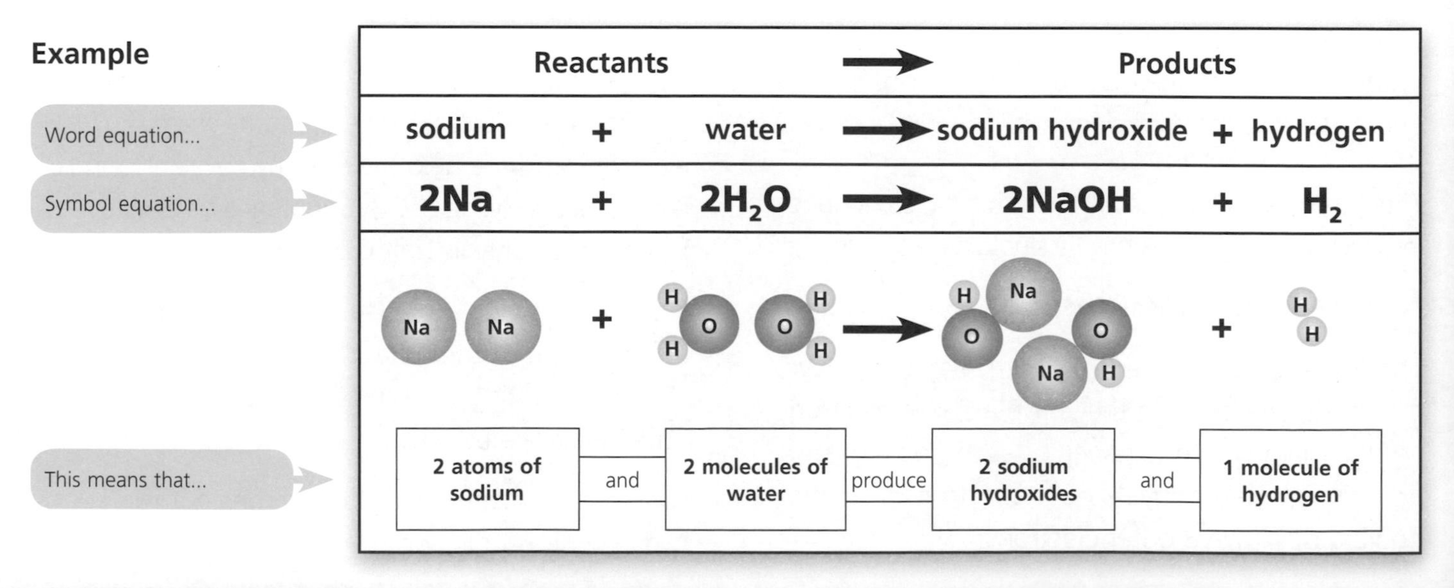

Writing Balanced Equations

Follow these steps to write a balanced equation:

1. Write a word equation for the chemical reaction.
2. Substitute formulae for the elements or compounds involved.
3. Balance the equation by writing numbers in front of the reactants and / or products.
4. Write the balanced symbol equation.

Example 1 – The reaction between magnesium and oxygen.

	Reactants			→	Products
1 Write a word equation	magnesium	+	oxygen	→	magnesium oxide
2 Substitute formulae	Mg	+	O_2	→	MgO
3 Balance the equation	Mg	+	O O	→	Mg O
	Mg	+	O O	→	Mg O Mg O
	Mg Mg	+	O O	→	Mg O Mg O
	Mg Mg	+	O O	→	Mg O Mg O
4 Write a balanced symbol equation	2Mg	+	O_2	→	2MgO

- There are two **O**s on the reactant side, but only one **O** on the product side. We need to add another **MgO** to the product side to balance the **O**s
- We now need to add another **Mg** on the reactant side to balance the **Mg**s
- There are two magnesium atoms and two oxygen atoms on each side – **it is balanced**.

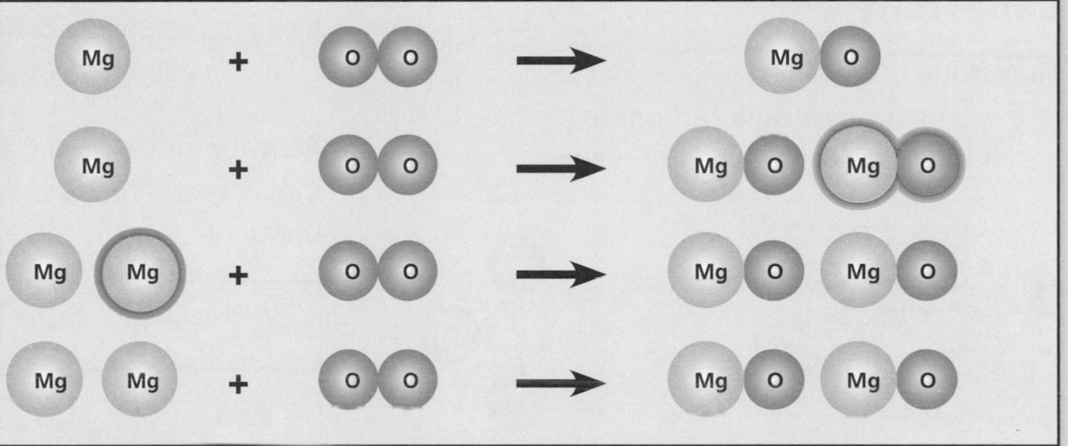

Example 2 – The production of ammonia.

	Reactants			→	Products
1 Write a word equation	nitrogen	+	hydrogen	→	ammonia
2 Substitute formulae	N_2	+	H_2	→	NH_3
3 Balance the equation	N N	+	H H	→	H H N H
	N N	+	H H	→	H H N H, H H N H
	N N	+	H H, H H, H H	→	H H N H, H H N H
	N N	+	H H, H H, H H	→	H H N H, H H N H
4 Write a balanced symbol equation	N_2	+	$3H_2$	→	$2NH_3$

- There are two **N**s on the reactant side, but only one **N** on the product side. We need to add another **NH_3** to the product side to balance the **N**s
- We now need to add two more **H_2**s on the reactant side to balance the **H_2**s
- There are two nitrogen atoms and six hydrogen atoms on each side – **it is balanced**.

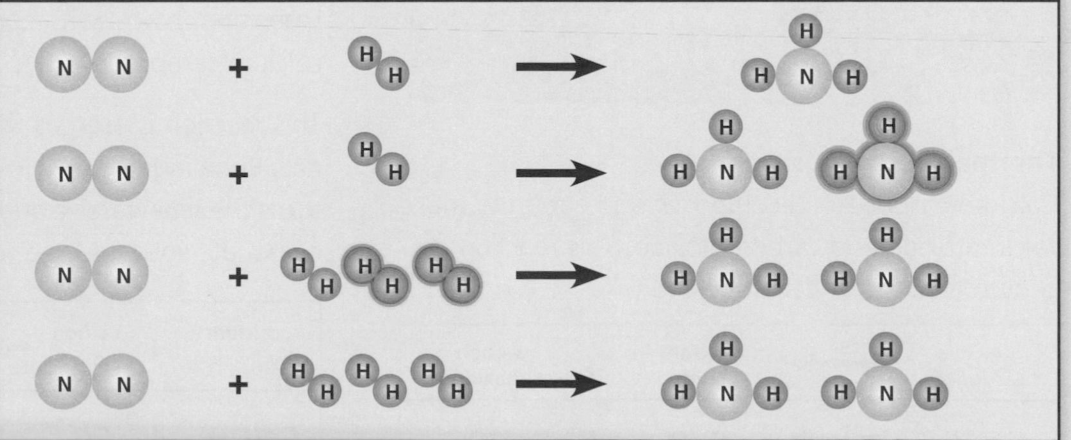

C1.2 Limestone and building materials

Rocks provide essential building materials. Limestone is a naturally occurring resource that provides a starting point for the manufacture of cement and concrete. To understand this, you need to know:

- why limestone is a useful resource
- how limestone is used to produce building materials.

Limestone

Limestone is a sedimentary rock that consists mainly of the compound **calcium carbonate** ($CaCO_3$). It is cheap, easy to obtain and has many uses.

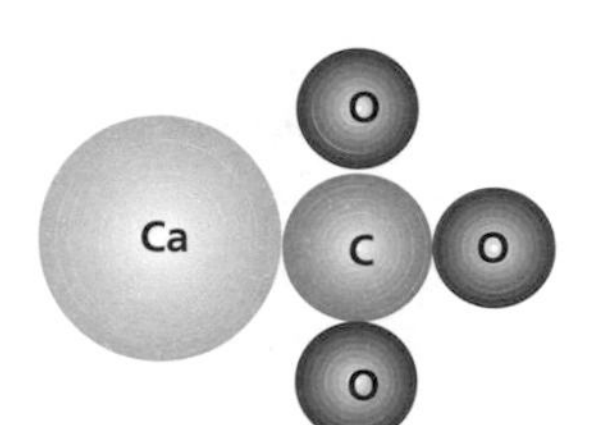

Building Material

Limestone can be **quarried** and cut into blocks, and used to build walls of houses in regions where limestone is plentiful.

Thermal Decomposition

Calcium carbonate decomposes on heating to make calcium oxide and carbon dioxide. This reaction is known as thermal decomposition.

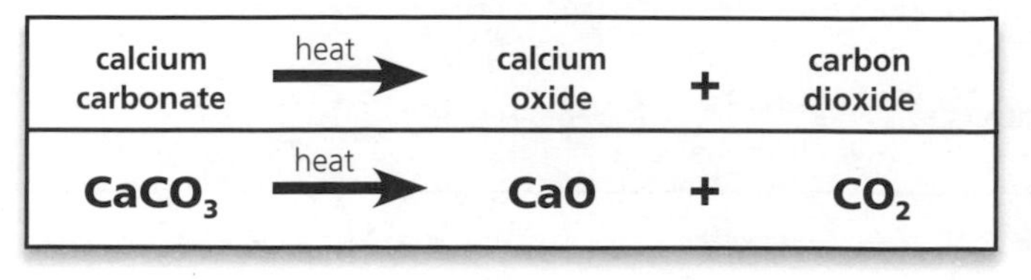

calcium carbonate	$\xrightarrow{\text{heat}}$	calcium oxide	+	carbon dioxide
$CaCO_3$	$\xrightarrow{\text{heat}}$	CaO	+	CO_2

Other metal carbonates undergo thermal decomposition on heating. They also produce a metal oxide and carbon dioxide gas. For example:

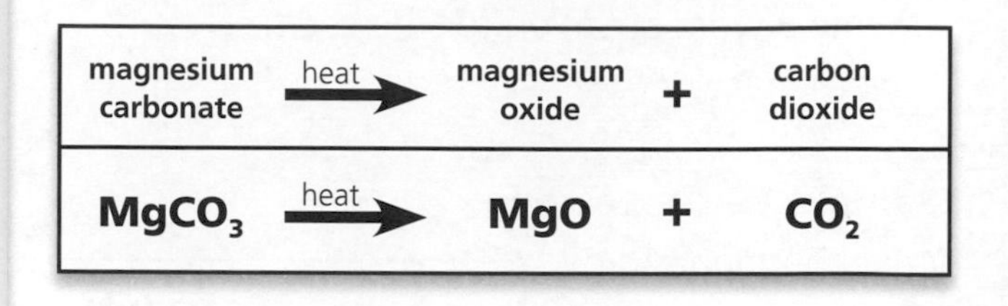

magnesium carbonate	$\xrightarrow{\text{heat}}$	magnesium oxide	+	carbon dioxide
$MgCO_3$	$\xrightarrow{\text{heat}}$	MgO	+	CO_2

N.B. The carbonates of other metals behave very similarly when they are heated.

Some Group 1 metal carbonates require higher temperatures than a Bunsen burner can provide in order to undergo thermal decomposition.

Reaction of Calcium Oxide

Calcium oxide reacts with water to form calcium hydroxide.

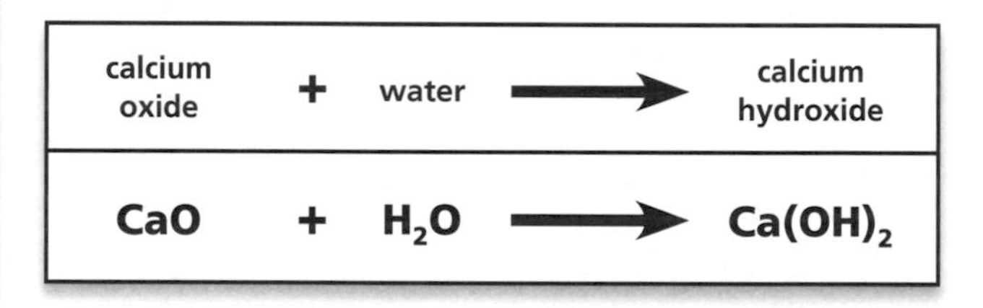

calcium oxide	+	water	$\rightarrow$	calcium hydroxide
CaO	+	H_2O	$\rightarrow$	$Ca(OH)_2$

Calcium hydroxide (like all metal hydroxides), is a strong alkali. It can be used to neutralise soils and lakes much faster than powdered limestone.

Reaction of Calcium Hydroxide

Calcium hydroxide solution (also known as limewater) reacts with carbon dioxide to form calcium carbonate.

This reaction is used as the test for carbon dioxide gas. Bubbling carbon dioxide gas through limewater turns the limewater cloudy (this is actually calcium carbonate solid you see forming).

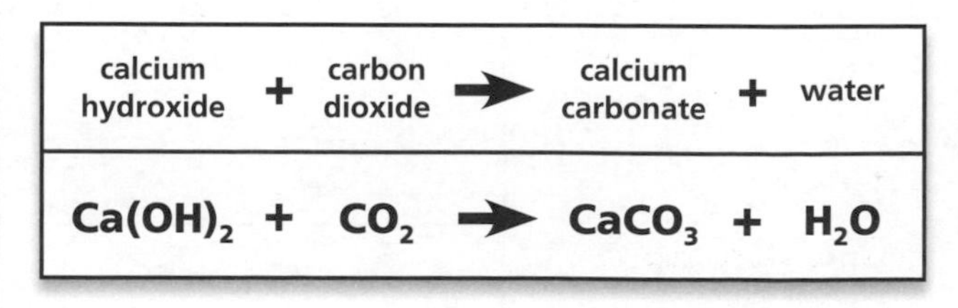

calcium hydroxide	+	carbon dioxide	$\rightarrow$	calcium carbonate	+	water
$Ca(OH)_2$	+	CO_2	$\rightarrow$	$CaCO_3$	+	H_2O

Limestone Cycle Summary

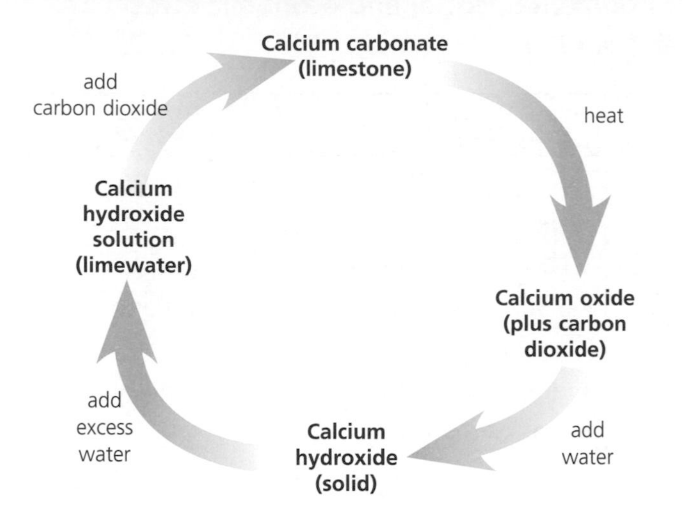

Cement, Mortar and Concrete

Powdered limestone and powdered clay are roasted in a rotary kiln to produce dry **cement**.

When the cement is mixed with sand and water it produces **mortar**, which is used to hold together bricks and stone during building.

When the cement is mixed with water, sand and gravel (crushed rock) a slow reaction takes place in which a hard, stone-like building material, called **concrete**, is produced.

Reaction of Metal Carbonates

Carbonates such as calcium carbonate (limestone) can react with acids to form a salt, water and carbon dioxide gas. All metal carbonates that you will meet in this unit behave in the same way.

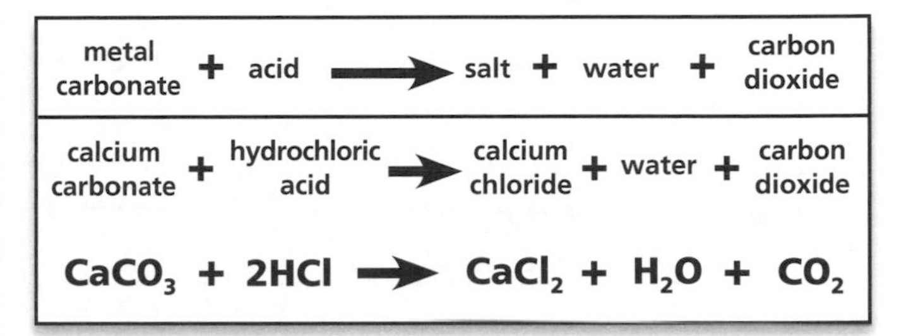

metal carbonate + acid ⟶ salt + water + carbon dioxide

calcium carbonate + hydrochloric acid ⟶ calcium chloride + water + carbon dioxide

$$CaCO_3 + 2HCl \longrightarrow CaCl_2 + H_2O + CO_2$$

This means that, over time, limestone (calcium carbonate) can become damaged by acid rain.

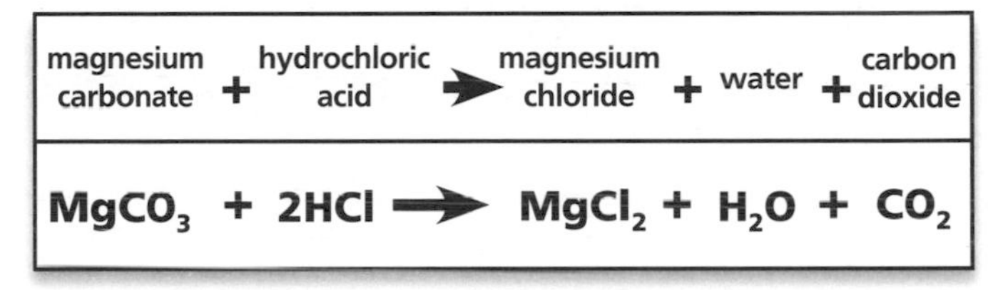

magnesium carbonate + hydrochloric acid ⟶ magnesium chloride + water + carbon dioxide

$$MgCO_3 + 2HCl \longrightarrow MgCl_2 + H_2O + CO_2$$

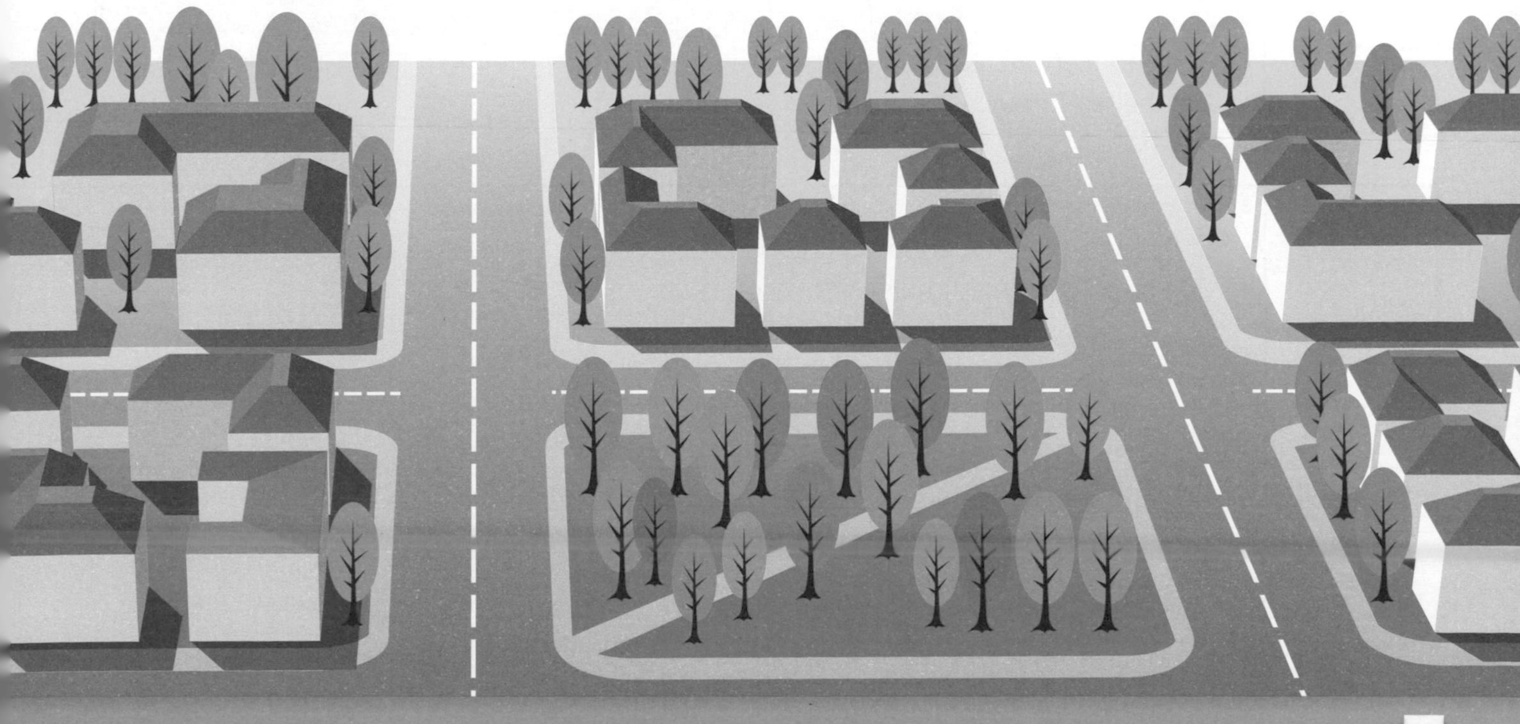

How Science Works

You need to be able to consider and evaluate the environmental, social and economic effects of exploiting limestone and producing building materials from it.

Advantages	Disadvantages
• Limestone is found naturally, so can be quarried relatively easily. • Using local stone to build new houses makes them 'fit in' with older houses. • Better roads will be built to cope with quarry traffic. • Creates more jobs locally. • Other industries (e.g. cement makers) will be attracted to the area, providing more job opportunities. • The quarry might invest in the local community in a bid to 'win over' the locals.	• Could be more expensive to quarry limestone than to use another building material. • Quarries destroy the landscape and the habitats of animals and birds. • Increased traffic to and from the quarries. • Noise pollution. • Health problems arising from the dust particles, e.g. asthma. • Reduced tourism in the area.

You need to be able to evaluate the advantages and disadvantages of using limestone, concrete and glass as building materials.

Material	Advantages	Disadvantages
Limestone	• Widely available. • Easy to cut. • Cheaper than many other building materials, e.g. marble. • Can be used to produce cement, concrete and glass.	• Susceptible to acid rain – the dilute acid dissolves the limestone very slowly, wearing it away.
Concrete	• Can be moulded into different shapes, e.g. panels and blocks which can be put together easily in buildings. • Quick and cheap way to construct buildings. • Does not corrode, so is a good alternative to metal. • Can be reinforced using steel bars so that it is safer and has a wider range of uses.	• Low tensile strength* and can crack and become dangerous, especially in high-rise buildings. • Looks unattractive.
Glass	• Transparent, therefore useful for windows and parts of buildings where natural light is wanted. • Can be toughened or made into safety glass. N.B. You don't need to know about glass in detail – but be aware that you may be given sufficient information to make comparisons with other materials	• Breaks easily. • Not always the cheapest or safest option. • Toughened and safety glass are expensive.

**Tensile strength refers to a material's ability to resist breaking when under tension.*

All of these materials are resistant to fire and rot, and for most building requirements are strong enough to resist attack from animals and insects, which make them a better choice than wood. However, there may be cheaper, safer and more aesthetically pleasing materials that are also suitable for the job.

C1.3 Metals and their uses

We use metals constantly in our everyday lives. Most metals are obtained from ores, which are naturally occurring rocks that provide an economic starting point for the manufacture of metals. Copper, iron, aluminium and titanium are examples of commonly used metals extracted from ores. Using more of these metals is creating a shortage of some ores, e.g. copper-rich ores, so new methods of extraction are being researched. To understand this, you need to know:

- what an ore is
- how we obtain different metals from their ores
- the properties and uses of different metals such as copper, iron, aluminium and titanium
- what an alloy is and why they are important
- the impact on the environment caused by the extraction of metals and methods being used to overcome this.

Ores

The Earth's crust contains many naturally occurring elements and compounds called **minerals**. A metal **ore** is a mineral or mixture of minerals from which economically viable amounts of pure metal can be **extracted**. This can change over time. After ores have been mined they may be concentrated before the metal is extracted and purified.

Extracting Metals from their Ores

The method of extraction depends on how reactive the metal is. Unreactive metals like gold exist as the metal itself (native metal). They are obtained through physical processes such as panning.

Most metals are found as **metal oxides** or compounds that can be easily changed into a metal oxide. To extract a metal from its oxide the oxygen must be removed by heating the oxide with another element in a chemical reaction. This process is called **reduction**.

Metals that are less reactive than carbon can be extracted from their oxides by heating them with carbon. (The carbon is a more reactive element, so it will displace the metal and form a compound with the oxygen.)

Metals that are more reactive than carbon are extracted from their ores using **electrolysis**. The ore is heated until it is liquid (molten) before the metal can be extracted.

Electrolysis requires large amounts of electricity and is, therefore, an expensive method to use. Reactive metals such as aluminium have to be extracted using electrolysis.

The Transition Metals

In the centre of the **Periodic Table**, between Group 2 and Group 3, is a block of metallic elements called the **transition metals**. These include iron, copper, platinum, mercury, chromium, titanium and zinc.

These metals are **hard** and mechanically **strong**. They have **high melting points** (except mercury – which is liquid at room temperature).

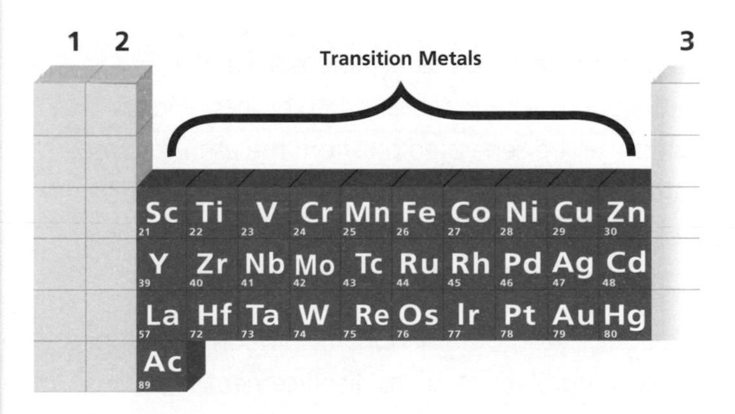

Transition metals, like all other metals, are **good conductors** of heat and electricity, and can also be easily bent or hammered into shape.

These properties make transition metals very useful as structural materials, and as electrical and thermal conductors.

Metals and their Uses

Copper

Copper is used for electrical wiring and plumbing because of its properties. Copper:

- is a good conductor of heat and electricity (good for use as electrical wires)
- is malleable (can be hammered into shape and bent), but is also hard enough to make pipes
- does not react with water (good for use as water pipes)
- can be drawn into wires (ductile).

Copper is a very valuable metal and can be extracted from copper-rich ores using a furnace. This process is known as smelting. The resultant copper is then purified using electrolysis. Because copper is such a versatile and useful material we are rapidly depleting our supplies. Therefore, the supply of copper-rich ores is limited.

To enable us to continue using copper, new ways of extracting the metal from low-grade ores are currently being researched. These new extraction methods include phytomining and bioleaching, and not only allow us to use low-grade ores but they also limit the environmental impact of more traditional mining methods.

Phytomining uses plants to absorb metal compounds. The plants are then burned and the metal can be separated out from the ash produced. **Bioleaching** uses bacteria to produce **leachate** solutions that contain metal compounds. The metal can then be extracted from this solution.

Copper is also obtained from solutions of copper salts using electrolysis or displacement reactions using the more reactive metal iron. Electrolysis involves the positive copper ions being attracted to the negative electrode. Electrolysis is dealt with more thoroughly in Unit C2 (page 69).

Aluminium and Titanium

Aluminium and titanium are both useful metals because they have a low density and are resistant to corrosion. Aluminium reacts with oxygen from the air and so a layer of aluminium oxide coats the metal, which prevents further corrosion. Aluminium is used in making drink cans, window frames and aeroplanes. Titanium is used in making aeroplanes, nuclear reactors and replacement hip joints.

Aluminium and titanium are more reactive than carbon and, therefore, cannot be extracted from their oxides using reduction. Aluminium is extracted using **electrolysis**, which makes it expensive.

Electrolysis is very complex – there are lots of different stages – and requires a large amount of energy. This makes it very expensive. So, we should recycle metals wherever possible to:

- save money and energy
- make sure we do not use up all the natural resources
- reduce the amount of mining because it is damaging the environment.

Iron

Iron is less reactive than carbon and can, therefore, be extracted from its ore by reduction using carbon in a blast furnace.

Molten iron obtained from a blast furnace contains roughly 96% iron along with 4% carbon and other metals. Because it is impure, the iron is very brittle and has limited uses. To produce **pure iron**, all the impurities would have to be removed.

Alloys

An alloy is a mixture of metals, or a metal and at least one other element. The added element disturbs the regular arrangement of the metal atoms so that the layers do not slide over each other so easily. Alloys are, therefore, usually stronger and harder than pure metals. Many of the metals we use everyday are alloys.

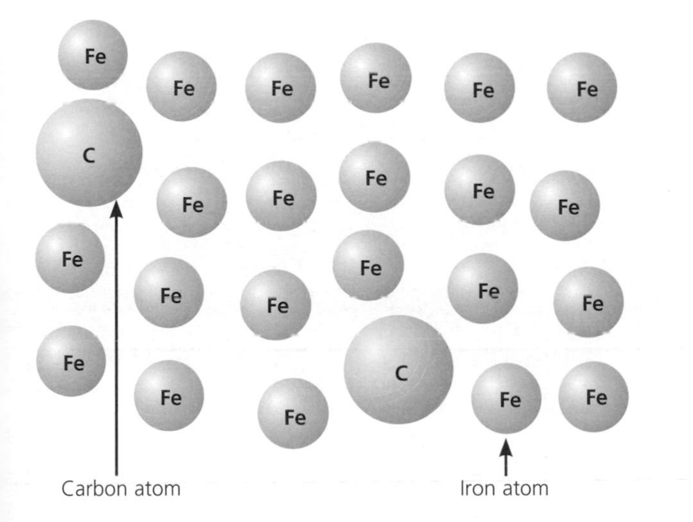

Pure copper, gold and aluminium are too soft for many uses. They are mixed with small amounts of similar metals to make them harder for everyday use.

Steel

Pure iron is soft, and iron from the blast furnace is brittle and easily corrodes (rusts), so most iron is converted into steel. Carbon is usually added to iron to make the steel (an alloy). However, some steels, such as stainless steel, are produced by adding other metals to the iron.

Most iron is converted into steel because this makes it more useful.

Alloys like steel are developed to have the necessary properties for a specific purpose. In steel, the amount of carbon and/or other elements determines its properties:
- steel with a high carbon content is hard and strong, e.g. for screwdrivers
- steel with a low carbon content is soft and easily shaped – mild steel (0.25% carbon) is easily pressed into shape, e.g. for cars
- steel that contains chromium and nickel is called stainless steel – it is hard and resistant to corrosion, e.g. for knives and forks.

Recycling and Environmental Impact

Metal ores are limited resources. The continued extraction and use of metals means that we are running out of some metals and, therefore, we need to recycle the metals we do have in order to preserve these limited resources. Extraction often requires the use of large amounts of energy, which is both expensive and damaging to the environment.

Extraction methods often lead to pollution or contamination of land by metals such as cadmium, nickel and cobalt. Plants like cabbages (brassicas) can be used to remove unwanted metals from the soil. The metals are then removed by taking the plants out of the soil.

Smart Alloys

Smart alloys belong to a group of materials that are being developed to meet the demands of modern engineering and manufacturing. These materials respond to changes in their environment, e.g. temperature, moisture, pH and electrical and magnetic fields.

Smart alloys (also called shape-memory alloys) remember their shape. They can be deformed, but will return to their original shape (usually when they are heated).

You need to be able to consider and evaluate the social, economic and environmental impacts of exploiting metal ores, of using metals and of recycling metals.

Example

The Lonsdale News, Tuesday 20 September 2011 17

Local Village Launches National Campaign

The villagers of Littlehampton are speaking out to show us all how we can help to reduce the damage being done to our environment.

A large metal extraction plant was built near Littlehampton 15 years ago and the village has suffered from the effects of the industry ever since. The plant has had a detrimental impact not only on the look of the area but also on its environment. The pollution has made its way into rivers and streams, and the local wildlife group reports that certain species are dwindling in number.

These bad effects are not limited to wildlife. The noise that comes from the factory is annoying, but more worrying is the dramatic increase in the number of local people who suffer from asthma as a result of the dust particles discharged into the air.

Now, after proposals to extend the plant have been revealed, villagers have joined together to recycle metals and want to encourage others to do the same. Campaign spokesman Bob Jeffries, 42, says, 'We know that metals are useful materials and that extraction needs to be done, but we hope to encourage people to recycle what they can. This should reduce the demand for newly extracted metals and remove the need for new plants. Not only will we be improving our standard of living but we will also be helping to reduce the pressure and impact on our environment'.

Recycling is a much better option because it uses less energy, which makes it a much cheaper process. The more times a material is recycled, the more cost-effective it becomes.

Councillors listened to the views of local residents and agreed to implement a recycling scheme. Five large recycling bins have been brought in to the car park at the local supermarket. Said councillor Cilla Jackson, 56, 'The scheme has had a much better response than we had hoped for; I just hope the enthusiasm for it is maintained.'

For information on how you can do your bit go to www.recyclenow.com

Method	Advantages	Disadvantages
Extracting metals	• Provides jobs and income locally. • Provides raw materials for industry. • Local facilities (e.g. roads) improved to cope with additional traffic.	• Destroys the landscape. • Leads to a reduction in tourism. • Noise and dust pollution. • Traffic problems.
Recycling metals	• Saves energy (e.g. less energy used to recycle aluminium than to electrolyse the ore it comes from). • Less pollution produced because fewer materials are sent to landfill sites. • Less pressure placed on the environment (the more material that is recycled, the less pressure there is to find new materials to mine, extract, etc.).	• Individual apathy. • Availability and collection of recycling facilities can make recycling difficult.

You need to be able to evaluate the advantages and disadvantages of using metals as structural materials and as smart materials.

Use of Metal	Advantages	Disadvantages
Structural material	• Hard, tough and strong. • Do not corrode easily. • Can be bent or hammered into shape. • Alloys of metals are harder than pure metals. • Relatively inexpensive at present.	• Iron is naturally very soft so needs to be mixed with other metals to form steel. • Conduct electricity and heat – this might not be what is wanted. • Some metals, particularly iron, can be corroded by water and other chemicals – this weakens and eventually wears away the metal. • The supply of metal ores from the Earth's crust is decreasing as more is extracted – eventually the supplies will run out.
Smart material	• Can be produced by mixing metals – have many advantageous properties over the original metals. • Good mechanical properties, e.g. strong and resist corrosion. • Can return to their original shape when heated (known as the shape-memory effect) – used in thermostats, coffee pots, hydraulic fittings. • More bendy than normal metals, therefore harder to damage. • Have new properties, such as pseudo-elasticity, which can be exploited in diverse ways such as in glasses frames, bra underwires and orthodontic arches. • Can be changed by passing an electrical current or a magnetic field through them or by heating. • Not much temperature change required (sometimes as little as 10°C) to change the structure.	• Expensive to manufacture. • Fatigue easily – a steel component can survive for around 100 times longer than a smart material under the same pressure.

C1.4 Crude oil and fuels

Crude oil is found trapped in porous rocks and consists of many different hydrocarbons. Crude oil can be separated by fractional distillation into different fractions, some of which are very useful as fuels. Other useful fuels can be produced from plant material (known as biofuels). Crude oil is a non-renewable resource whereas biofuels are a renewable resource. To understand this, you need to know:

- the difference between a compound and a mixture
- how crude oil can be fractionally distilled to produce fuels
- that most fuels contain carbon and hydrogen, and sometimes sulfur
- what is produced when fuels burn and the impact that this has on the environment
- what alkanes are, their general formula and how to represent them.

Crude Oil

Crude oil is a mixture containing many different compounds. Most of these compounds are **hydrocarbons** (they contain only the atoms carbon and hydrogen). Mixtures contain different elements or compounds that are not chemically combined and, therefore, each part of the mixture keeps its own chemical properties.

Because crude oil is a mixture we can separate it using physical methods such as distillation. This separates crude oil into different fractions. Each fraction is made up of similar size hydrocarbons that have similar boiling points.

To separate crude oil using fractional distillation, the oil is vaporised (by heating), pumped into a fractionating column and the vapour is allowed to condense at different temperatures. The fractions, each of which contains hydrocarbons with a similar number of carbon atoms, are collected.

The smaller hydrocarbons (those with fewer carbon and hydrogen atoms) are more useful as fuels. This is because the properties of a hydrocarbon depend on the size of the carbon chain. The longer the carbon chain:

- the higher the boiling point
- the more viscous it is (i.e. it flows less easily)
- the less volatile it is (i.e. it does not vaporise/turn into a gas easily)
- the less flammable it is (i.e. it is much harder to ignite)
- the more soot it produces when it burns.

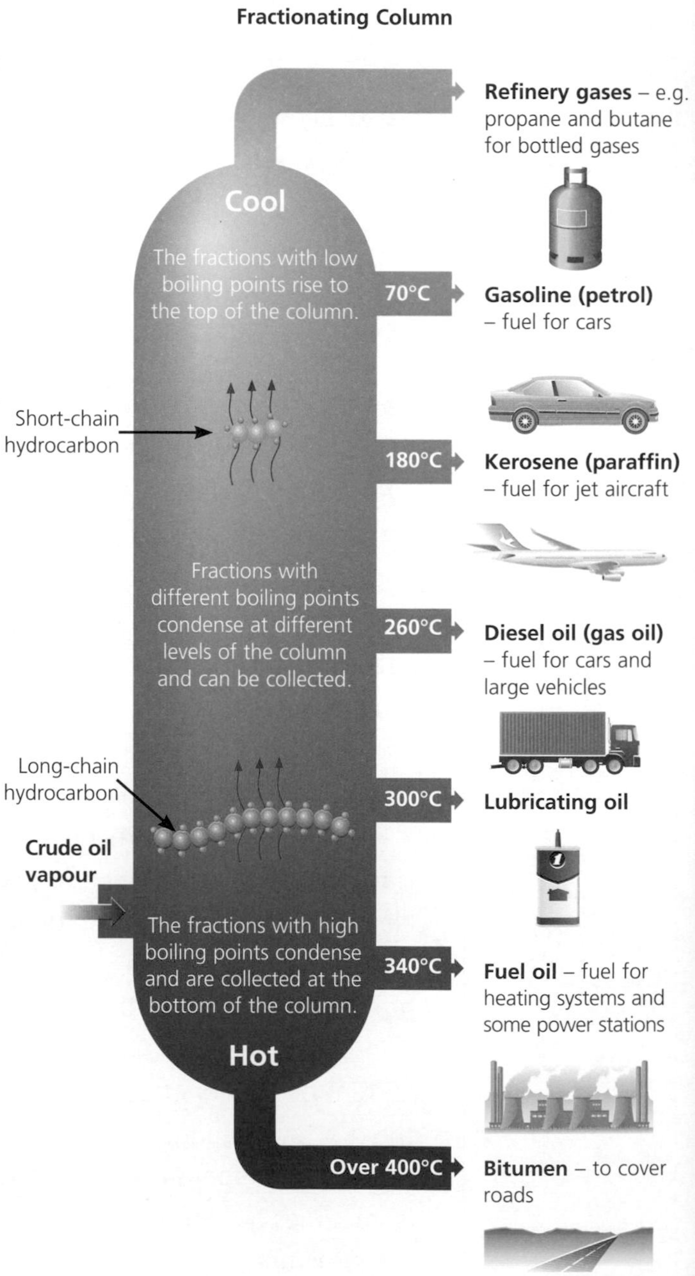

Fractionating Column

Refinery gases – e.g. propane and butane for bottled gases

Cool

The fractions with low boiling points rise to the top of the column.

70°C **Gasoline (petrol)** – fuel for cars

Short-chain hydrocarbon

180°C **Kerosene (paraffin)** – fuel for jet aircraft

Fractions with different boiling points condense at different levels of the column and can be collected.

260°C **Diesel oil (gas oil)** – fuel for cars and large vehicles

Long-chain hydrocarbon

300°C **Lubricating oil**

Crude oil vapour

The fractions with high boiling points condense and are collected at the bottom of the column.

340°C **Fuel oil** – fuel for heating systems and some power stations

Hot

Over 400°C **Bitumen** – to cover roads

Alkanes

Most of the hydrocarbons in crude oil are saturated molecules called alkanes. Each of the carbons in an alkane molecule has four single bonds to either other carbon atoms or hydrogen atoms. All of the carbon–carbon bonds are single covalent bonds so we say the hydrocarbon is **saturated**.

Alkanes have the general formula C_nH_{2n+2}. Alkanes can be represented by structural formula and displayed formula. The table shows the first three alkanes in the series.

Name	Formula	Displayed Formula
Methane	CH_4	H I H – C – H I H
Ethane	C_2H_6	H H I I H – C – C – H I I H H
Propane	C_3H_8	H H H I I I H – C – C – C – H I I I H H H

Butane has 4 carbon atoms (C_4H_{10}).

Covalent bonds are shown in the above structures using a straight line (–).

Hydrocarbon Fuels

Fuels usually contain carbon and often hydrogen, and when burned they produce energy (in the form of heat). This is known as a **combustion reaction**.

Combustion reactions are actually oxidation reactions because the carbon is oxidised (gains oxygen) to form carbon dioxide and the hydrogen is oxidised to form water vapour.

Combustion reactions can produce different gases depending on whether the reaction is complete or partial. This depends on the amount of oxygen.

Short-chain hydrocarbons are very useful as fuels. When burned, hydrocarbon fuels release gases into the atmosphere. The gases include carbon dioxide, water vapour and possibly carbon monoxide. Sulfur is often present in the fuel as an impurity and so sulfur dioxide is often produced. Burning fuels can also lead to the production of nitrogen oxides and carbon particles (soot). Nitrogen oxides are formed at high temperatures.

Complete combustion of carbon produces carbon dioxide (CO_2) whereas partial or incomplete combustion produces carbon monoxide (CO) and solid particles that may contain unburned hydrocarbons and carbon particles (soot).

The gases and solid particles released by the combustion of fuels cause environmental problems:

- Carbon dioxide (CO_2) causes 'global warming' due to the greenhouse effect.
- Sulfur dioxide (SO_2) causes acid rain. This problem can be reduced by removing sulfur from the fuel before burning, or SO_2 from the waste gases after burning – however this can be costly. Power stations remove SO_2 from the waste gases produced when combustion occurs to reduce the amount of pollution they emit.
- Carbon particles (soot) cause global dimming (a reduction in the amount of sunlight reaching the Earth's surface).
- Nitrogen oxides can also cause acid rain.

Biofuels

Biofuels are produced from plant materials and include ethanol and biodiesel. There are advantages and disadvantages in using these types of fuels rather than fuels produced from crude oil. (See page 26).

You need to be able to consider and evaluate the social, economic and environmental impacts of the uses of fuels.

Example

Forget Fossil Fuels...

When hydrocarbons are burned they release harmful waste gases into the air. Burning fossil fuels has a considerable impact on the environment and its inhabitants. The solution is to use less fossil fuel by using alternative energy resources, using existing resources more efficiently and by making changes to lifestyles, such as car-sharing to reduce fuel consumption, etc.

...Switch to Sugar!

Brazilian motorists have been converting their petrol-guzzling cars so they can be powered by ethanol. Ethanol, better known as grain alcohol, is easily distilled from sugar cane and is a cheap alternative to petrol. This new use for sugar cane has greatly affected farmers – never has there been such a high demand for sugar! This in turn has helped to push up the price of sugar to an all-time high. Scientists hope that the use of sugar cane as a viable alternative to petrol will grow, because it burns more cleanly than usual fuels and produces less of the harmful gas, carbon monoxide. However, it is not all good news. Alcohol releases less energy than petrol when it burns and it can be a health risk to filling station attendants.

You need to be able to evaluate developments in the production and uses of better fuels, for example, ethanol and hydrogen.

Example

Is Rocket Fuel the Way Forward?

Experts are carrying out research to find out if hydrogen gas, which is currently used as rocket fuel, could be an answer to our pollution problem. Hydrogen can be produced by passing an electric current through water, and when it burns it releases a lot of energy. Unlike other fuels which produce harmful gases when burned, hydrogen produces only water vapour, which does not pollute the atmosphere. A major consideration is the cost involved because the production of hydrogen requires a lot of electricity. It could also be quite risky because hydrogen burns explosively so it needs to be stored under special conditions.

Fuel	Advantages	Disadvantages
Fossil fuel	• Power stations provide jobs. • Provides energy for homes and industry. • Does not take up much space.	• Produces pollutants. • Causes global warming due to the greenhouse effect. • Non-renewable source so is in danger of running out.
Ethanol	• Does not affect the performance of a car. • Can save money. • Can be made from renewable resources. • Less carbon emissions. • Less carbon monoxide produced.	• Need to pay out to convert engines. • Much more sugar will need to be grown to meet demand. • Price of sugar is likely to rise due to increased demand. • Produces less energy than petrol when it burns.
Hydrogen	• No harmful gases produced, only water vapour which does not harm the environment.	• Expensive to produce (requires lots of energy). • Difficult to store and transport safely.

C1.5 Other useful substances from crude oil

Fractions produced from the distillation of crude oil can be cracked to produce alkenes such as ethene. Alkenes can be used to make polymers and ethanol. To understand this, you need to know:

- how hydrocarbons are cracked
- how fuels and the starting materials for polymers can be produced from cracking
- how polymers are formed and their properties and uses
- how ethanol is produced from both renewable and non-renewable resources.

Cracking Hydrocarbons

Long-chain hydrocarbons can be cracked (broken down) into shorter chains using heat. When long-chain hydrocarbons are cracked, a short-chain alkane and alkene are produced. The short-chain alkane is more useful as a fuel and the short-chain alkene can be used to make polymers.

Hydrocarbons are cracked by heating them until they vaporise, then passing the vapour over a heated catalyst, where a thermal decomposition reaction takes place.

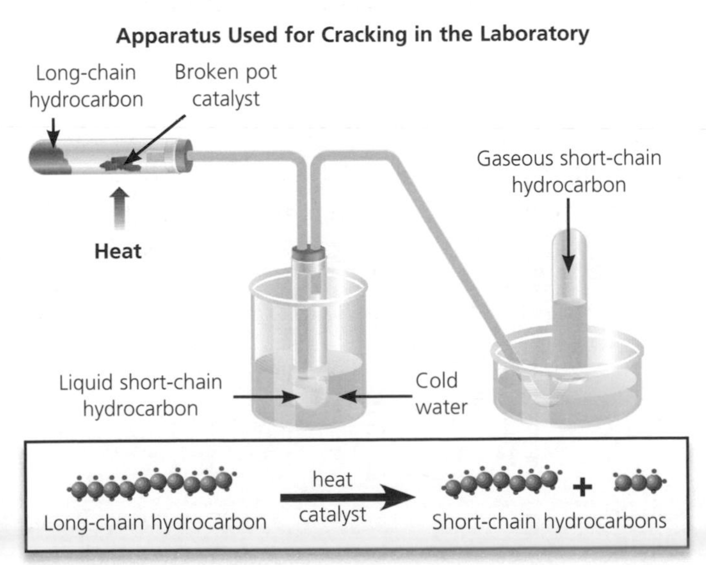

Hydrocarbons can also be cracked by mixing them with steam and heating them to very high temperatures.

Alkenes (Unsaturated Hydrocarbons)

We have already seen that carbon atoms can form single bonds with other atoms, but they can also form double bonds. Some of the products of cracking are hydrocarbon molecules with at least one double bond; this is an **unsaturated** hydrocarbon and it is known as an alkene. Alkenes always have names ending in 'ene'.

The general formula for alkenes is C_nH_{2n}. The simplest alkene is ethene, C_2H_4, which is made up of 4 hydrogen atoms and 2 carbon atoms. As you can see in the diagram below, ethene contains one double carbon–carbon bond (often represented by '=').

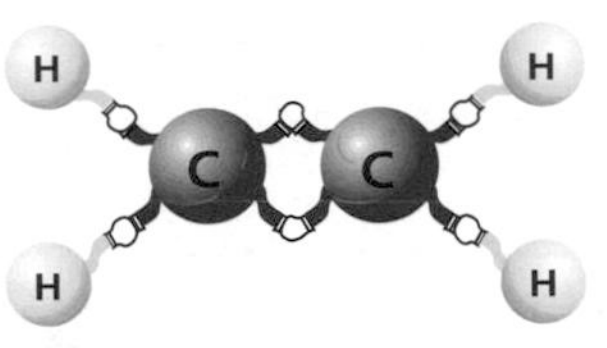

Here is another way of representing alkenes:

Name	Formula	Displayed Formula
Ethene	C_2H_4	H H \ / C=C / \ H H
Propene	C_3H_6	H H \ | C=C−C−H / | | H H H

Not all the carbon atoms are linked to 4 other atoms; a double carbon–carbon bond is present instead. Because they are unsaturated, they are useful for making other molecules, especially **polymers**.

When alkenes react with orange bromine water it turns colourless. This reaction can be used to distinguish between alkanes and alkenes because alkanes do not react with bromine water in this way.

Polymerisation

Because alkenes are unsaturated (have a double bond), they are very reactive. When small alkene molecules (monomers) join together to form long-chain molecules (polymers), it is called **polymerisation**.

The properties of polymers depend on what they are made from and the conditions under which they are made. The materials commonly called plastics are all synthetic polymers. They are produced commercially on a very large scale and have a wide range of properties and uses. Polymers and plastics were first discovered in the early 20th century.

Making Poly(ethene) from Ethene

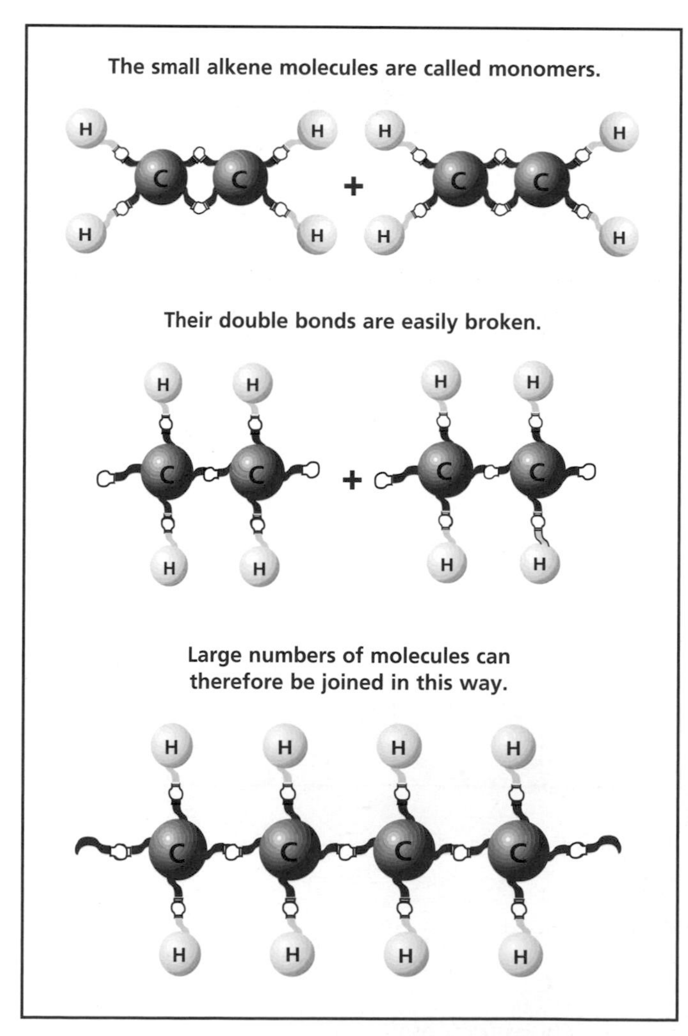

The resulting long-chain molecule is a polymer – in this case poly(ethene), often called polythene. Poly(propene) can be made in a similar way.

Representing Polymerisation

A more convenient way of representing polymerisation is:

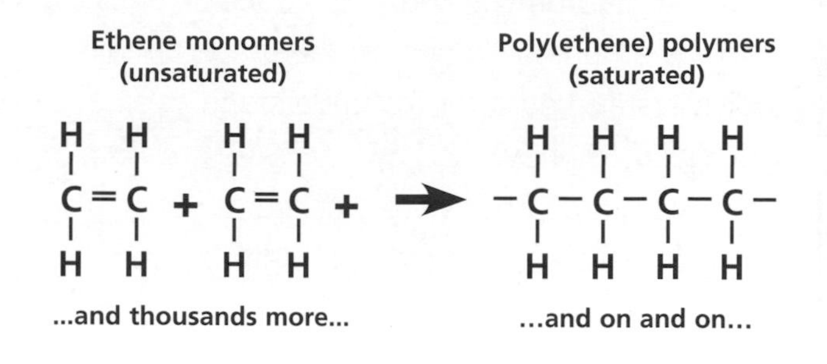

General Equation for Polymerisation

This equation can be used to represent the formation of any simple polymer:

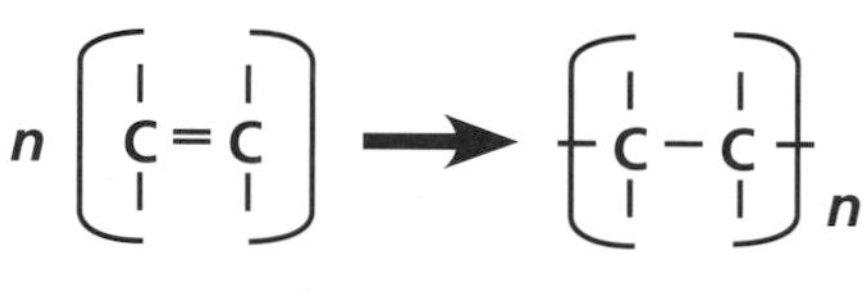

where n is a very large number

For example, if we take n molecules of propene we can produce poly(propene), which is used to make crates and ropes:

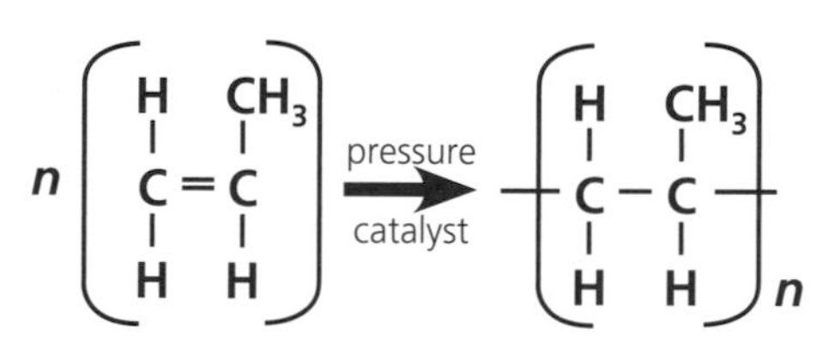

And n molecules of chloroethene can produce polychloroethene (also known as polyvinyl chloride or PVC):

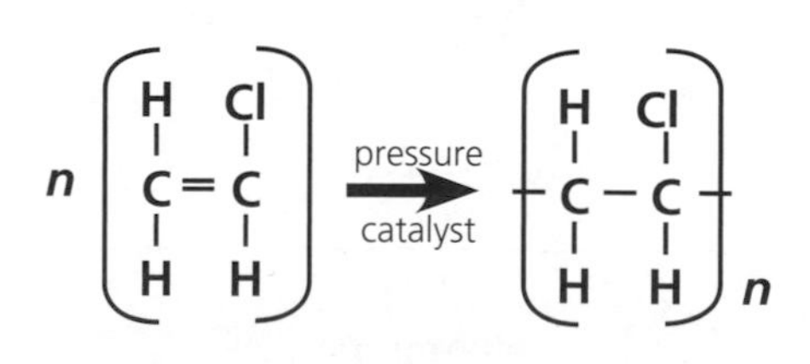

Polymers are classified by the reactions by which they are formed. These are examples of **addition** polymers, formed by addition polymerisation.

Other Useful Substances from Crude Oil

Uses of Polymers

Specific polymers can have different uses, for example:

- **Polyvinyl chloride (PVC)** can be used to make waterproof items and drain pipes and can also be used as an electrical insulator.
- **Polystyrene** is used to make casings for electrical appliances and it can be expanded to make protective packaging.
- **Poly(ethene)** is commonly used to make plastic bags and bottles.
- **Poly(propene)** can be used to make crates and ropes.

Polymers have many useful applications and new uses are being developed.

Polymers and composites are widely used in medicine and dentistry:

- Implantable materials are used for hard and soft tissue surgery, replacing and fusing damaged bone and cartilage.
- Hard-wearing anti-bacterial dental cements, coatings and fillers have been produced.
- Hydrogels can be used as wound dressings.
- Silicone hydrogel contact lenses have been developed over the last few years. Research has shown that people who wear this type of contact lens have a 5% lower risk of developing severe eye infections.

Polymers and composites can be used to coat fabrics with a waterproof layer. Smart materials, including shape-memory polymers, are also increasingly more common.

Disposing of Plastics

Because plastic is such a versatile material and it is cheap and easy to produce, we tend to generate a large amount of plastic waste.

The main problem with most plastics is that they are not biodegradable. This means that if they are left as litter or buried in landfill sites they cannot be broken down by microbes and will therefore not decompose and rot away. This leads to a build-up of plastic waste in landfill sites.

Plastics can also be burned but this produces air pollution, including toxic fumes from some plastics.

One way in which scientists are trying to overcome the problem with the disposal of plastics is to make biodegradable polymers. An example of this is a polymer made from cornstarch.

Plastics are recycled in order to reduce the amount sent to landfill.

Making Alcohol from Ethene

Ethanol is an **alcohol**. It can be produced by reacting steam with ethene at a moderately high temperature and pressure in the presence of a catalyst, phosphoric acid.

ethene + **steam** $\xrightarrow{\text{phosphoric acid}}$ **ethanol**

Ethanol can be used as:

- a solvent
- a fuel.

Producing ethanol from ethene means using a non-renewable resource (crude oil) and this is one of the main disadvantages of this method.

Making Alcohol from Sugar

Ethanol can also be produced from sugar, which is a renewable resource. This process is called fermentation and is carried out in solution using yeast at temperatures of about 37°C.

The process produces ethanol and carbon dioxide.

Fermentation is used to make alcoholic drinks but it can also be used to make ethanol to burn as a fuel.

You need to be able to evaluate the social and economic advantages and disadvantages of using products from crude oil as fuels, or as raw materials for plastic and other chemicals.

Crude oil is one of our most important natural resources. It is hard to imagine what our lives would be like without the products that we can get from crude oil. Transport would come to a standstill, there would be no more plastics and detergents, and the pharmaceutical industry would not be able to get essential raw materials, so medicines would run out.

Crude oil can be used to make tough, lightweight, waterproof and breathable fabrics for clothes; paint for cars; dyes; packaging and communication equipment. However, it is important to weigh up the advantages of the products we can get from crude oil against the disadvantages of using it as a raw material, particularly since it is a limited resource.

Advantages of Crude Oil

- The oil industry provides jobs.
- The fractions of crude oil have many uses.
- Provides raw materials for industry.
- Provides fuel for transport.

Disadvantages of Crude Oil

- Oil spills damage the environment.
- Can cause air pollution.
- Increases global warming.
- Produces non-biodegradable material.

You need to be able to consider and evaluate the social, economic and environmental impacts of the uses, disposal and recycling of polymers.

Plastics (polymers) are everywhere. There are a wide range of polymers with different, highly useful physical properties – some polymers are flexible, others are rigid; some have a low density, whereas others are very dense. They can be transparent or opaque. They are waterproof and resistant to corrosion and can be used as a protective layer.

However, although polymers are relatively cheap to produce, the cost to society and the environment has to be considered. Pollution, and its effects on residents who live near polymer-producing factories, is a major issue. And the problem of disposal of polymers once they have been used needs to be addressed because burning them produces harmful and sometimes toxic gases.

Advantages of Polymers

- Cheap to make.
- Many uses because of their different properties.
- Provide jobs in factories that make the polymers and the products.
- Some polymers can be recycled, melted down and made into something else, which saves valuable natural resources.
- If polymers are used instead of wood, fewer trees will have to be cut down.

Disadvantages of Polymers

- People do not like to live near polymer-producing industrial works.
- Some people think plastic products look cheap compared with natural materials.
- Made from oil, a non-renewable resource.
- Most plastics are not biodegradable so there is a problem of how to get rid of them.
- Give off toxic fumes when they burn.
- Sorting types of polymers for recycling can be expensive.

You need to be able to evaluate the advantages and disadvantages of making ethanol from renewable and non-renewable sources.

Using Non-renewable Sources

Ethanol can be produced by reacting steam with ethene at a moderately high temperature and pressure in the presence of the catalyst, phosphoric acid.

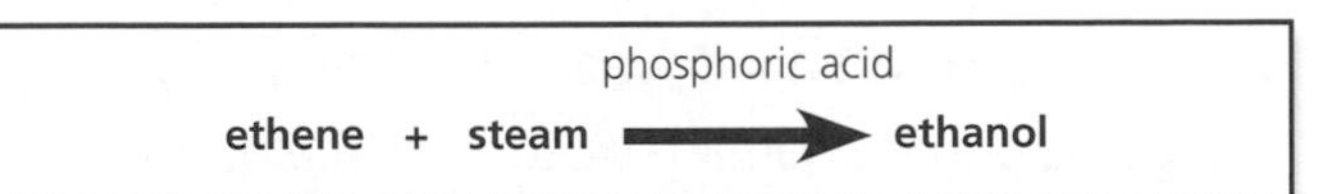

Using Renewable Sources

Ethanol can also be produced by the fermentation of sugars. Water and yeast are mixed with the raw materials at just above room temperature. Enzymes, which are biological catalysts found in the yeast, react with the sugars to form ethanol and carbon dioxide. The carbon dioxide is allowed to escape from the reaction vessel, but air is prevented from entering it. The ethanol is separated from the reaction mixture by fractional distillation when the reaction is over.

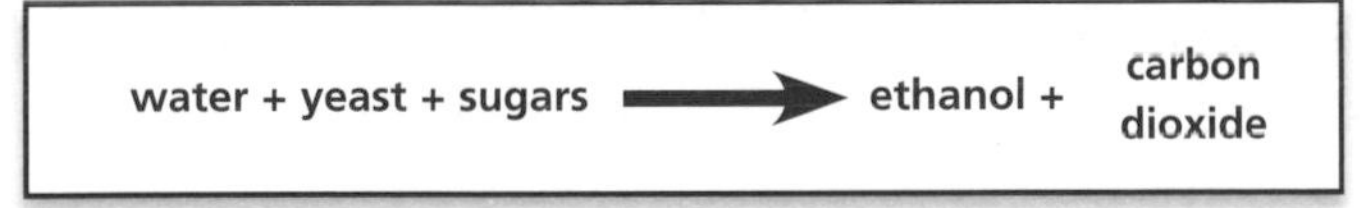

One problem with the production of ethanol is that it can be oxidised by air (in certain conditions) to produce ethanoic acid. The presence of ethanoic acid results in alcoholic drinks turning sour.

Method	Advantages	Disadvantages
Reacting ethene with steam	• Fast rate of production. • High-quality ethanol produced. • Can be produced continuously. • Best method for making large quantities.	• Uses non-renewable raw material (crude oil). • High temperatures needed.
Fermentation	• Renewable raw material (sugar). • Fairly high-quality ethanol produced after fractional distillation. • Best method for making small quantities. • Lower temperatures needed.	• Slow rate of production. • Is produced in batches. • Ethanol not as pure as that produced by reacting ethene with steam.

C1.6 Plant oils and their uses

Many plants produce oils that can be extracted and used for many purposes. Vegetable oils can be hardened to make margarine or to make biodiesel fuel. Emulsions can be made and have a number of uses. To understand this, you need to know:

- how oils are extracted from plants
- the properties and uses of oils
- what emulsions are.

Getting Oil from Plants

Many plants produce fruit, seeds and nuts that are rich in **oils**, which can be extracted and made into consumer products. Some common examples you might find in the food you eat are:

- sunflower oil
- olive oil
- oilseed rape
- palm kernel oil.

Oil can be extracted from plant materials by pressing (crushing) them or by distillation. This removes the water and other impurities from the plant material.

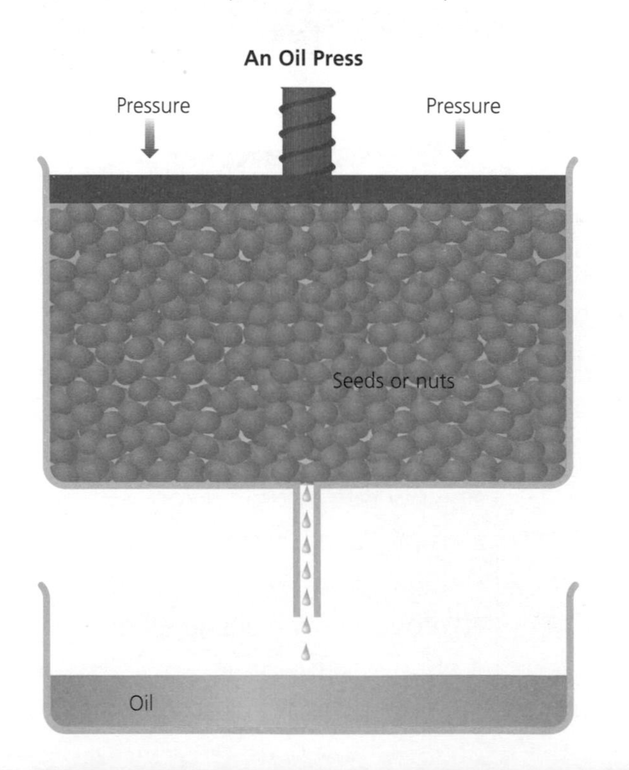

Vegetable Oils

Vegetable oils are important **foods** and **fuels** because they provide nutrients and a lot of energy. Some vehicles can now be converted to use vegetable oils as their fuel instead of petrol or diesel.

Vegetable oils contain double carbon–carbon bonds, so they are described as **unsaturated**. They can be detected using bromine water. They react with the orange-coloured bromine water to decolourise it. The bromine becomes part of the compound, by breaking the double bond. For example:

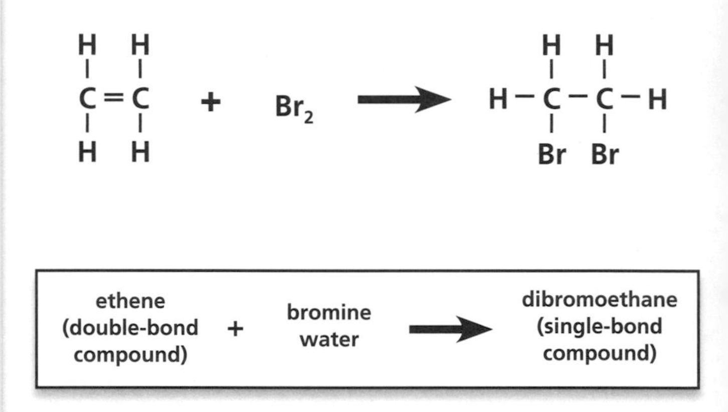

They will react in the same way with iodine (I_2).

Vegetable oils are used to cook foods because they boil at higher temperatures than water and, therefore, allow food to be cooked at higher temperatures. This changes both the texture and taste of food. Cooking at higher temperatures in oil means that food is cooked more quickly. It also increases the energy that the food produces when eaten because it is coated in oil.

Emulsions

Oils do not dissolve in water. Oil and water have different densities. A mixture of oil and water is called an **emulsion**. If oil and water are mixed thoroughly, droplets of oil can be seen dispersed in the water.

Emulsions are thicker than oil or water and have a better texture, appearance and coating ability. They have many uses, e.g. salad dressing and ice cream.

If you mix some olive oil and vinegar together you can make a salad dressing. However, it does not stay mixed for very long because the water particles in the vinegar clump together and the oil particles clump together. The mixture can be seen to separate into two layers.

You can make a salad dressing last longer by adding some mustard to it before shaking the mixture up. This stops the separate layers forming. The mustard is an **emulsifying agent**.

Many different emulsifying agents are used in the manufacture of food to stop vegetable oils and water forming separate layers.

Emulsifiers are used to stop oil and water separating, for example egg yolks are used as emulsifiers in mayonnaise.

Emulsifiers work because different parts of the molecule are attracted to the two different liquids. The 'heads' of the emulsifiers are attracted to the water and the 'tails' are attracted to the oil.

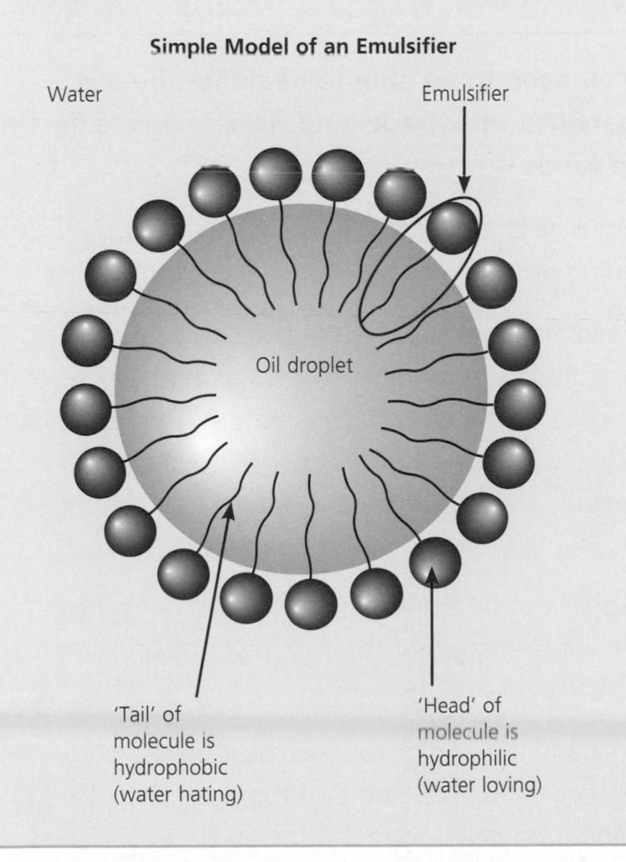

The Manufacture of Margarine

As a general guide, the more double carbon–carbon bonds present in a substance, the lower its melting point. This means that unsaturated fats, e.g. vegetable oils, tend to have melting points below room temperature and are called oils.

For some purposes you might need a solid fat, for example, to spread on your bread or to use to make cakes and pastries. You can raise the melting point of an oil to above room temperature by removing some or all of the double carbon–carbon bonds.

When ethene and hydrogen are heated together in the presence of a nickel catalyst, a reaction takes place that removes the double carbon–carbon bonds to produce ethane. This process is called **hydrogenation** and is used to convert unsaturated hydrocarbons into more saturated hydrocarbons.

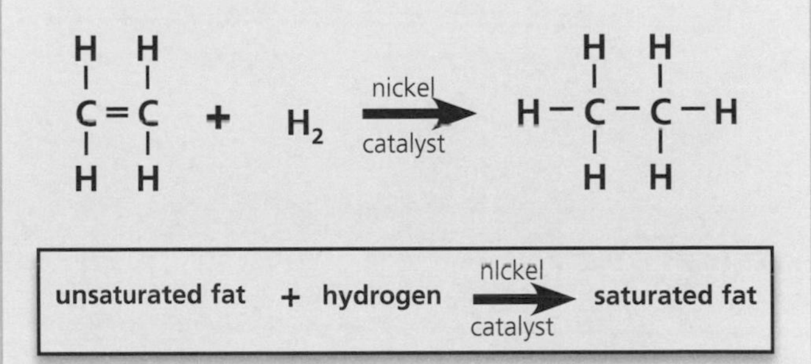

Margarine is manufactured from unsaturated vegetable oils like sunflower oil. The oil is reacted with hydrogen, at a temperature of around 60°C in the presence of a nickel catalyst and some of the double bonds are hydrogenated. Removing more double bonds makes the margarine harder. Unfortunately, hydrogenation of vegetable oils also makes them less healthy to eat.

You need to be able to evaluate the effects of using vegetable oils in food and the impacts on diet and health.

Oils have many uses. However, the amount of saturated fats we consume needs to be carefully controlled to reduce the risk of heart disease.

Vegetable oils are a healthy alternative to using fats derived from animals because they contain monounsaturated fats that can lower blood cholesterol levels, and they contain no cholesterol.

However, it is important to remember that you should not consume too much of any oil because this would not lead to a healthy balanced diet. Where possible, unsaturated fats such as olive oil should be used to reduce the health risks. However, using saturated fats occasionally would not be too bad for your health.

You need to be able to evaluate the benefits, drawbacks and risks of using vegetable oil to produce fuels.

Cars can now be converted so that they can run on vegetable oil instead of diesel. Although in many ways this is a more environmentally friendly option, it has not yet become widespread, and only a few cars have been converted.

Advantages of Vegetable Oils as Fuels
• Cheaper than diesel or petrol. • Fewer pollutant gases produced – virtually carbon neutral. • No change to performance of car. • Renewable source.
Disadvantages of Vegetable Oils as Fuels
• High cost of conversion kit. • Unpleasant smell. • Inconvenience of filling up your car – vegetable oil currently not an option at petrol stations. • Increased demand may put up prices for food made using vegetable oil. • Large areas of land needed to grow oil crops.

You need to be able to evaluate the use, benefits, drawbacks and risks of emulsifiers in foods.

Emulsifiers are additives used in foods to produce emulsions.

Emulsifiers and stabilisers (e.g. lecithin, E322) are used to mix ingredients that would normally separate, in order to give a consistent texture. They are used along with other additives to give food a better texture, taste, shelf-life and appearance. Emulsifiers make the fat more difficult to detect in foods.

Additives are very useful, but the benefits need to be weighed against the risks involved in using them in food.

Any additive added to food must be shown on the list of ingredients. Some of the additives added to food have been given E-numbers.

C1.7 Changes in the Earth and its atmosphere

The Earth is made up of four layers and is surrounded by an atmosphere that both protects us and provides the conditions needed for life. The atmosphere has changed drastically since the Earth was first formed, but has remained constant for the last 200 million years. Recent human activities have caused further changes in the atmosphere. To understand this, you need to know:

- the structure of the Earth
- what tectonic plates are, how they move and what this causes
- what the atmosphere consists of and how it was formed from the early atmosphere.

Structure of the Earth

The Earth is nearly spherical and has a layered structure that consists of:

- a thin crust – thickness varies between 10km and 100km
- a mantle – extends almost halfway to the centre and has all the properties of a solid even though it does flow very slowly
- an inner and outer core (made of nickel and iron) – over half of the Earth's radius. The outer core is liquid and the inner core solid.

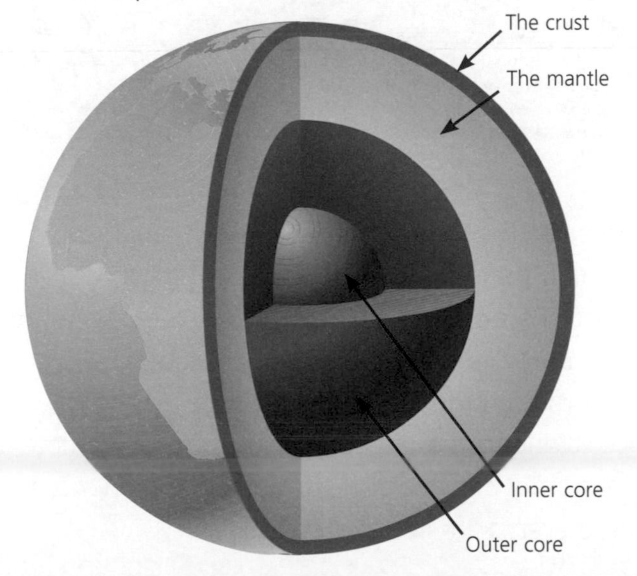

The average density of the Earth is much greater than the average density of the rocks that form the crust, because the interior is made of a different, denser material than that of the crust.

Although there does not seem to be much going on, the Earth and its crust are very dynamic. Rocks at the Earth's surface are continually being broken up, reformed and changed in an ongoing cycle of events known as the rock cycle. It is just that the changes take place over a very long time.

Tectonic Theory

At one time people used to believe that features on the Earth's surface were caused by shrinkage as the Earth cooled, following its formation. However, as scientists have found out more about the Earth, this theory has now been rejected.

A long time ago, scientists noticed that the east coast of South America and the west coast of Africa have:

- similar patterns of rocks, which contain fossils of the same plants and animals, e.g. the Mesosaurus
- closely matching coastlines.

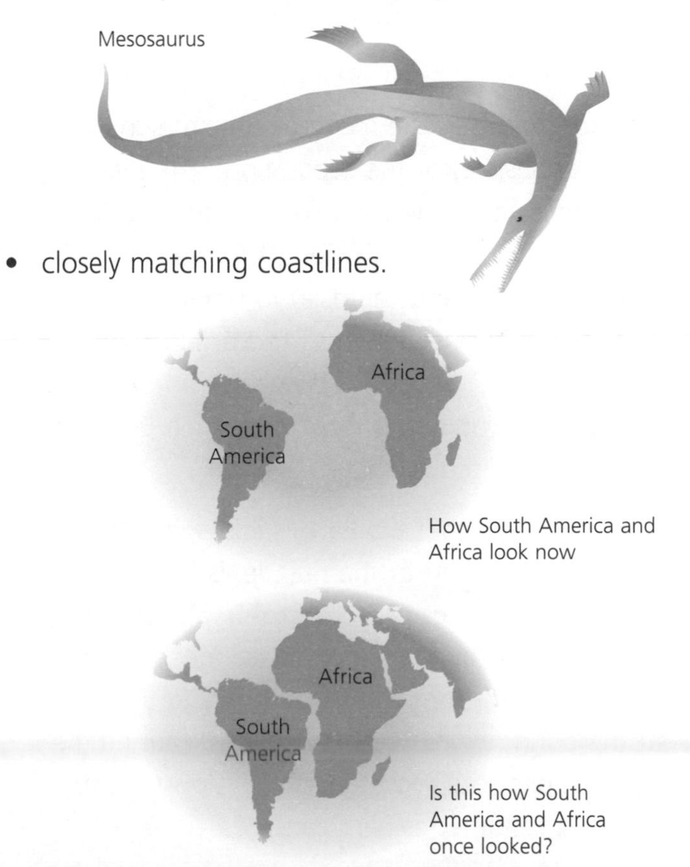

How South America and Africa look now

Is this how South America and Africa once looked?

Tectonic Theory (cont)

This evidence led Alfred Wegener to propose that, even though they are now separated by thousands of kilometres of ocean, South America and Africa had at one time been part of a single land mass.

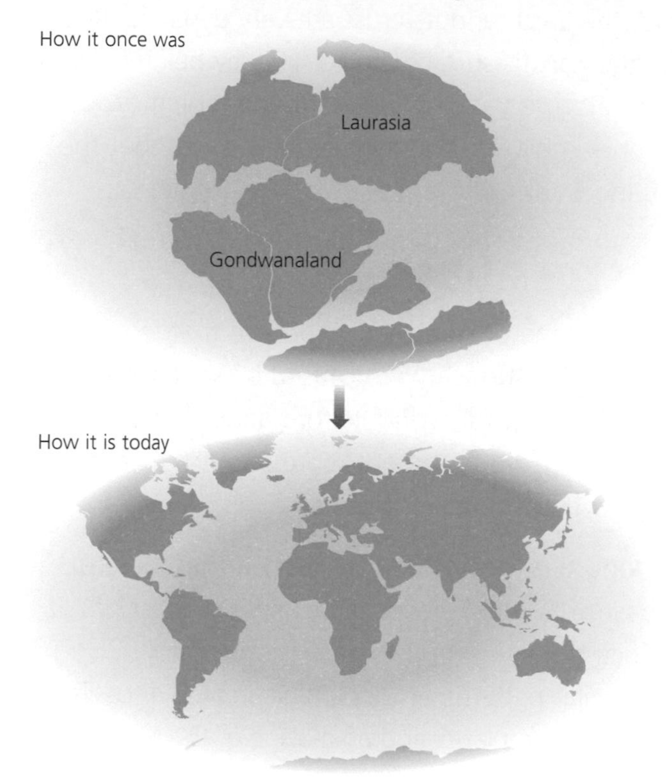

He proposed that the movement of the crust was responsible for the separation of the land (continental drift), which explains the movement of the continents from where they were (as Gondwanaland and Laurasia) to how they look today. This is known as **tectonic theory**. Unfortunately, Wegener was unable to explain *how* the crust moved and it took more than 50 years for scientists to discover this.

We now know that the Earth's lithosphere (the crust and the upper part of the mantle) is 'cracked' into several large pieces called **tectonic plates**. Intense heat, released by radioactive decay deep in the Earth, causes hot molten rock to rise to the surface at the boundary between the plates, causing the tectonic plates to move apart very slowly, at speeds of a few centimetres per year.

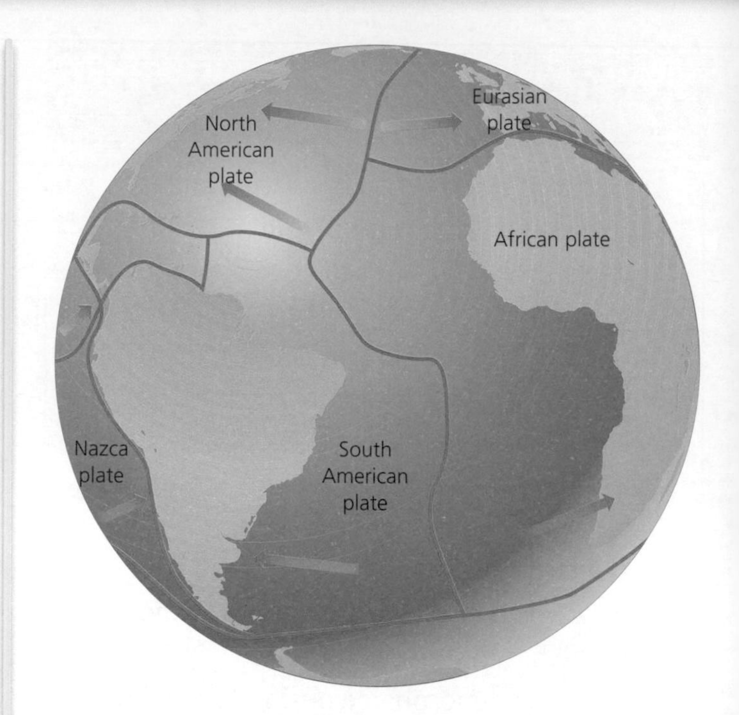

In convection in a gas or a liquid, the matter rises as it is heated, then as it gets further away from the heat source it cools and sinks down again. The same happens in the mantle. The hot molten rock rises to the surface, creating new crust. The older crust, which is cooler, then sinks down where the convection current starts to fall. This causes the land masses on these plates to move slowly across the globe.

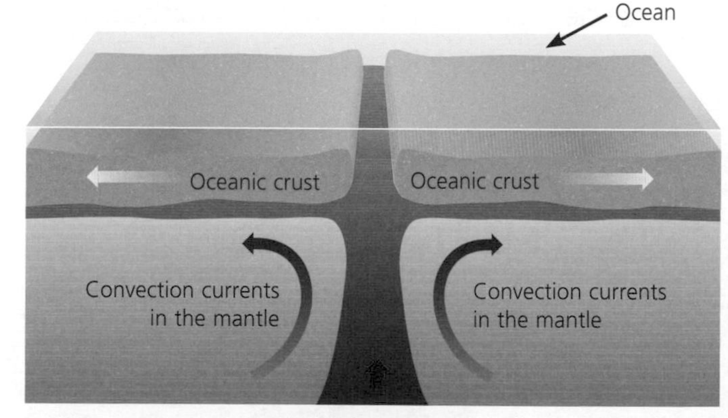

Although the movements are usually small and gradual, they can sometimes be sudden and disastrous. Earthquakes and volcanic eruptions are common occurrences at plate boundaries. As yet, scientists cannot predict *when* these events will occur, due to the difficulty in making appropriate measurements, but at least they do know *where* these events are likely to occur.

The Earth's Atmosphere

Since the formation of the Earth 4.6 billion years ago the atmosphere has changed a lot. The timescale, however, is enormous because one billion years is one thousand million (1 000 000 000) years!

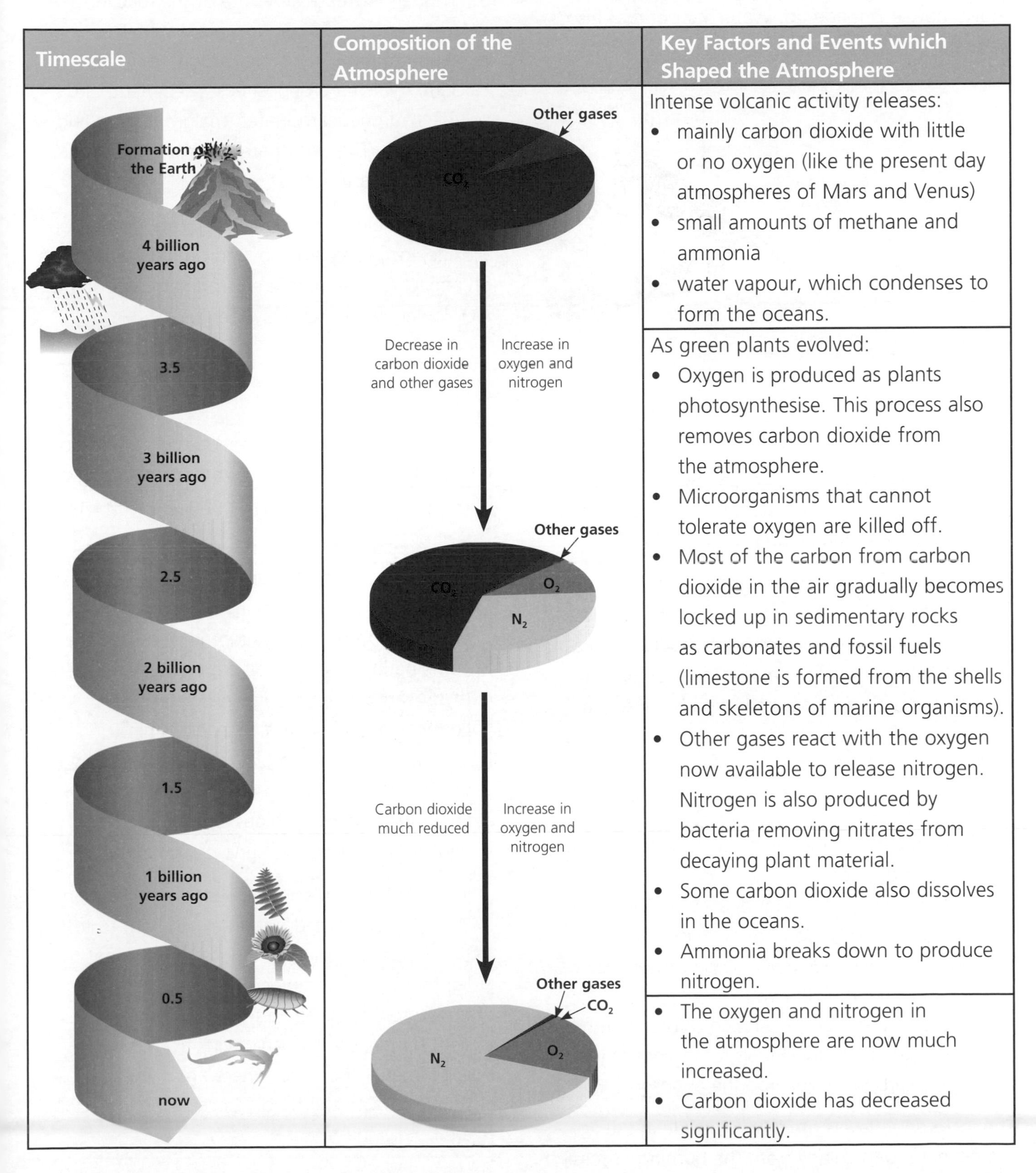

Timescale	Composition of the Atmosphere	Key Factors and Events which Shaped the Atmosphere
		Intense volcanic activity releases: • mainly carbon dioxide with little or no oxygen (like the present day atmospheres of Mars and Venus) • small amounts of methane and ammonia • water vapour, which condenses to form the oceans.
		As green plants evolved: • Oxygen is produced as plants photosynthesise. This process also removes carbon dioxide from the atmosphere. • Microorganisms that cannot tolerate oxygen are killed off. • Most of the carbon from carbon dioxide in the air gradually becomes locked up in sedimentary rocks as carbonates and fossil fuels (limestone is formed from the shells and skeletons of marine organisms). • Other gases react with the oxygen now available to release nitrogen. Nitrogen is also produced by bacteria removing nitrates from decaying plant material. • Some carbon dioxide also dissolves in the oceans. • Ammonia breaks down to produce nitrogen.
		• The oxygen and nitrogen in the atmosphere are now much increased. • Carbon dioxide has decreased significantly.

Composition of the Atmosphere

Our atmosphere has been more or less the same for about 200 million years. The pie chart below shows how it is made up. Water vapour may also be present in varying quantities (0–3%).

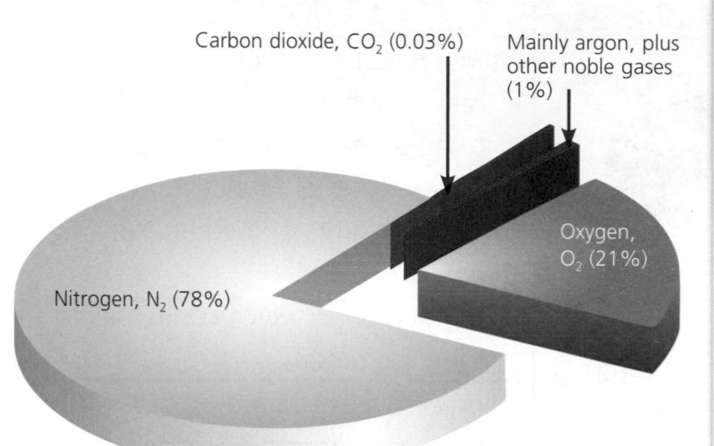

The noble gases (in Group 0 of the Periodic Table) are all chemically unreactive gases and are used in light bulbs and electric discharge tubes. Helium is much less dense than air and is used in balloons.

HT Air is a mixture containing different gases. These gases are used as a source of raw materials in a variety of industrial processes. To separate the gases we use fractional distillation because they all have different boiling points.

Changes to the Atmosphere

The level of carbon dioxide in the atmosphere today is increasing due to:

- volcanic activity – geological activity moves carbonate rocks deep into the Earth. During volcanic activity they may release carbon dioxide back into the atmosphere.
- burning of fossil fuels – burning carbon, which has been locked up in fossil fuels for millions of years, releases carbon dioxide into the atmosphere.

The increase in levels of carbon dioxide in the atmosphere, particularly from the burning of fossil fuels, is thought to cause global warming.

The level of carbon dioxide in the atmosphere is reduced by the reaction between carbon dioxide and sea water. Increased carbon dioxide in the atmosphere increases the rate of the reaction between carbon dioxide and sea water. This reaction produces insoluble carbonates (mainly calcium), which are deposited as sediment, and soluble hydrogencarbonates (mainly calcium and magnesium). The carbonates form sedimentary rocks in the Earth's crust.

Although the oceans act as a reservoir for carbon dioxide, increased levels of carbon dioxide do have an impact on the marine environment. Dissolving CO_2 lowers the pH of seawater.

HT Origin of Life

There are many different theories of how life on Earth began. At the moment there is no single accepted theory of how life originated on Earth.

One possible theory is known as the **Primordial Soup Theory**. This theory suggests that life on Earth began in an ocean or pond when a form of energy (lightning) caused the combination of chemicals from the atmosphere (hydrocarbons, ammonia and water) to combine and make amino acids. Amino acids are the building blocks of proteins. The primordial soup was an environment where, when the first living things were formed, they had all they needed to survive and reproduce.

In the 1950s two scientists, Stanley Miller and Harold Urey, carried out an experiment to test this theory. By using electricity to represent the lightning, they combined the gases thought to be present in the Earth's early atmosphere. They did generate amino acids from this experiment – however, there are many problems with the theory because amino acids are not living material.

You need to be able to explain why the theory of crustal movement (continental drift) was not generally accepted for many years after it was proposed.

About 200 years ago, most geologists thought that the Earth had gone through a period of being extremely hot and consequently had dried out and contracted, or shrunk, as it cooled. Features of the Earth, such as mountain ranges, were thought to have been wrinkles that formed in the Earth's crust as it shrank.

At this point, insufficient data had been collected to show that the continents were in fact moving, and nobody produced any evidence to contradict the theory of the Earth shrinking until the early 1900s.

Then Alfred Wegener studied certain features of the Earth (see p.35–36), which prompted him to propose his theory of continental drift in 1915. This theory proposed that the Earth is made up of plates that have moved slowly apart. Most geologists at the time said that this theory was impossible, although a few did support Wegener.

In the 1950s scientists were able to investigate the ocean floor and found new evidence to support Wegener's theory. They discovered that although he was wrong about some aspects, the basis for his theory was correct.

By the 1960s, geologists were convinced by the theory of continental drift and can now use it to explain many geological features and occurrences caused by moving tectonic plates. Evidence now shows that the sea floor is spreading outwards and convection currents in the mantle cause movement of the crust.

You need to be able to explain why scientists cannot accurately predict when earthquakes and volcanic eruptions will occur.

To understand how earthquakes and volcanic eruptions occur, we need to consider the movement of the tectonic plates. They can stay in the same position for some time, resisting a build up of pressure, and then when the pressure becomes too great they can suddenly move.

However, it is impossible to predict exactly when this will happen because the plates do not move in regular patterns. Scientists can measure the strain in underground rocks to see if they can calculate when an earthquake is likely to happen, but they are unlikely to be able to give an exact forecast.

A volcano erupts when molten rock rises up into the spaces between the rocks near the surface. Scientists have instruments which can identify these changes, and therefore warn of imminent eruptions. However, sometimes the molten rock cools, so the magma does not reach the surface and the volcano does not erupt. So, other factors which are hard to predict can affect whether a volcano erupts or not.

Therefore, despite having very sophisticated equipment which monitors volcanic activity and areas prone to earthquakes, scientists cannot always predict exactly when they might happen.

You need to be able to explain and evaluate the effects of human activities on the atmosphere.

Most human activities, especially those used to create heat and energy, can produce pollutants that are harmful to the atmosphere. Some of these activities and their impact on the atmosphere are listed below.

The burning of fossil fuels create the gases sulfur dioxide and carbon dioxide. Sulfur dioxide contributes to the formation of acid rain, which can erode buildings and add acid to lakes and the soil.

The carbon dioxide content of the air used to be roughly constant (0.03%) but it has been increased by the growth in population, which raises energy requirements.

Deforestation means less photosynthesis takes place. An increase in the level of carbon dioxide is believed to be responsible for global warming and climate change.

The combustion of petrol and diesel allows the reaction of nitrogen and oxygen at the very high temperatures in car engines to make oxides of nitrogen, which are pollutants. Carbon monoxide is produced by the incomplete combustion of fuels. It is a poisonous gas which is eventually oxidised to carbon dioxide, adding to global warming.

So, what can we do to reduce the pollutants? We need to create energy to heat homes, power our cars, etc., but we also need to look at the effect that various fuels have on the environment and consider alternative methods of producing energy in order to limit the impact we have on the planet.

- Cars fitted with catalytic converters reduce the production of carbon monoxide and oxides of nitrogen.
- Alternative forms of energy can reduce pollution, e.g. wind farms and hydroelectricity.
- Power stations can use fuels that reduce atmospheric pollution, e.g. nuclear power stations.

The European Union and the United Kingdom have made laws to control the pollution levels. These levels need to be regulated and monitored, and every country needs to control their pollution levels.

1. **a)** Draw the electronic structure for a chlorine atom. **(1 mark)**

   b) Explain why chlorine is in Group 7 of the Periodic Table. **(1 mark)**

2. **a)** What is meant by the term 'ore'? **(1 mark)**

   b) Metals are very useful in everyday life. Use the words in the box to complete the sentences below (you may not need to use all the words). **(3 marks)**

electrolysis	**less**	**more**	**oxidation**	**reduction**

Metals can be extracted from their ores using a process called .... This involves removing the oxygen from the metal oxide by heating it with another element. Metals that are .... reactive than carbon can be extracted from their ores by being heated with carbon. Metals that are more reactive than carbon are extracted from their ores by a process called ....

3. Crude oil is a mixture of different hydrocarbons. It is a very useful raw material. Most of the hydrocarbons in crude oil are alkanes.

   a) Which elements are found in a hydrocarbon? **(1 mark)**

   b) What is the general formula for an alkane? **(1 mark)**

   c) Long-chain alkanes are less useful than short-chain alkanes. Long-chain alkanes can be broken down to make more useful materials.

   i) What are the two conditions required for this reaction to occur? **(2 marks)**

   ii) Give the name of this reaction. **(1 mark)**

   iii) What type of compound is produced in this reaction along with a shorter-chain alkane? **(1 mark)**

   iv) Explain a chemical test to distinguish between the short-chain alkane and the other product of cracking (**part iii**). In your answer you should give the name of any chemical used and what observation you would make. **(2 marks)**

HT

4. Balance these symbol equations:

   a) $Na + O_2 \longrightarrow Na_2O$ **(1 mark)**

   b) $H_2SO_4 + NaOH \longrightarrow Na_2SO_4 + H_2O$ **(1 mark)**

C2 The Structure of Substances

C2.1 Structure and bonding

The arrangement of electrons in atoms can be used to explain what happens when elements react. To understand this, you need to know:

- how chemical bonds are formed
- how ionic, metallic and covalent bonds are formed (including how to represent these bonds using diagrams).

C2.2 How structure influences the properties and uses of substances

Atoms are held together by different types of bonds, which give rise to different structures that have different properties. Chemical bonds are strong and hard to break, whereas forces between molecules are weaker and easier to break. Nanomaterials are very small materials with new properties. To understand this, you need to know:

- that substances have different boiling and melting points depending on their structure
- why some substances can conduct electricity
- why giant and simple structures have different properties
- about the development of nanoscience
- reasons for the different properties of polymers.

Bonding

When atoms react together, chemical bonds are formed. These bonds occur when atoms **transfer** or **share** electrons from the outermost (highest) occupied energy level (shell). Atoms transfer or share electrons in order to gain a full outer shell. This gives atoms the electronic structure of a noble gas. When two or more different atoms form a chemical bond a compound is formed.

Compounds are substances in which two or more elements are chemically combined.

Types of Bonds

When atoms gain or lose electrons (by transferring electrons from one atom to another) **ions** are formed. Metal atoms **lose electrons** to form positive ions and non-metal atoms **gain electrons** to form negative ions. Atoms will gain or lose enough electrons in order to achieve the electronic structure of a noble gas.

An ionic bond is the electrostatic attraction between positive metal ions and negative non-metal ions.

When non-metal atoms share electrons, a covalent bond is formed.

Metallic bonds are the attraction between delocalised electrons and positive metal ions. They occur between metal atoms only.

Ionic Bonds

An ionic bond is made between a metal atom and a non-metal atom and involves a **transfer** of electrons from one atom to the other to form electrically charged **ions**, each of which has a complete outermost energy level or shell. This means that ions have the electronic structure of a noble gas. Atoms that **lose electrons** become **positively charged** ions; atoms that **gain electrons** become **negatively charged** ions.

Atoms in Groups 1–3 will lose electrons, but atoms in Groups 6–7 will gain electrons. The number of outer electrons an atom has is equal to its group number (for Groups 1–7).

Group	Gain or Lose Electrons	Charge on Ion Formed
1	Lose 1 electron	+1, e.g. Na^+
2	Lose 2 electrons	+2, e.g. Ca^{2+}
3	Lose 3 electrons	+3, e.g. Al^{3+}
5	Gain 3 electrons	–3, e.g. N^{3-}
6	Gain 2 electrons	–2, e.g. O^{2-}
7	Gain 1 electron	–1, e.g. Cl^-

Elements in Group 1 will react with non-metals by donating an electron to form an ionic compound in which the Group 1 metal has a single positive charge, e.g. Na^+.

Elements in Group 7 will react with metals by gaining an electron to form an ionic compound in which the Group 7 ion has a single negative charge, e.g. Cl^-.

Ionic compounds are giant structures of ions held together by strong forces of attraction between oppositely charged ions that act in all directions. This is called **ionic bonding**. Ionic compounds have high melting and boiling points.

Ionic compounds are made up of ions but have no charge overall – so the total positive charge must equal the total negative charge in any sample of any ionic compound.

For example sodium chloride is made up of Na^+ ions and Cl^- ions so there must be equal numbers of them, and its formula is NaCl. Sodium oxide is made up of Na^+ ions and O^{2-} ions so there must be twice as many Na^+ ions as O^{2-} ions, and its formula must be Na_2O.

You should be able to see that the formula of magnesium chloride (Mg^{2+} and Cl^-) is $MgCl_2$.

Example 1 – Sodium and chlorine bond ionically to form sodium chloride, NaCl. The sodium (Na) atom has 1 electron in its outer shell which is transferred to the chlorine (Cl) atom so they both have 8 electrons in their outer shells. The atoms become ions Na^+ and Cl^- and the compound formed is sodium chloride, NaCl.

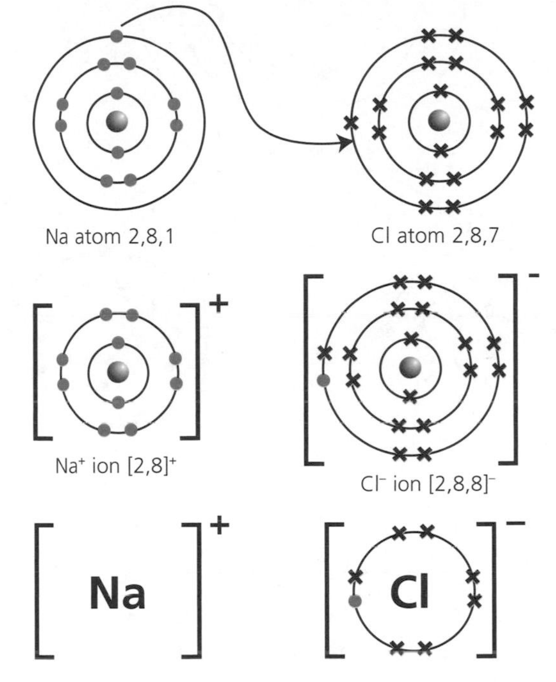

When showing ionic bonding you must make sure that one atom has its electrons shown as 'crosses', the other as 'dots'. This is to show where the electrons in the bond have come from. You can see from the diagrams that chlorine has gained one electron from sodium (the dot).

You only actually need to show the outer electrons in each diagram about ionic bonding. Therefore, the metal ends up having no outer electrons around it whereas the non-metal always has a full outer shell made up of both 'dots' and 'crosses' to show the electrons it had originally and those it gained from the metal.

Example 2 – Calcium and chlorine bond ionically to form calcium chloride, $CaCl_2$. The calcium (Ca) atom has 2 electrons in its outer shell and a chlorine (Cl) atom only accepts 1 electron – therefore, 2 Cl atoms are needed to give all 3 atoms 8 electrons in their outer shell. The atoms become ions Ca^{2+}, Cl^- and Cl^- and the compound formed is calcium chloride, $CaCl_2$.

Ionic Bonds (cont)

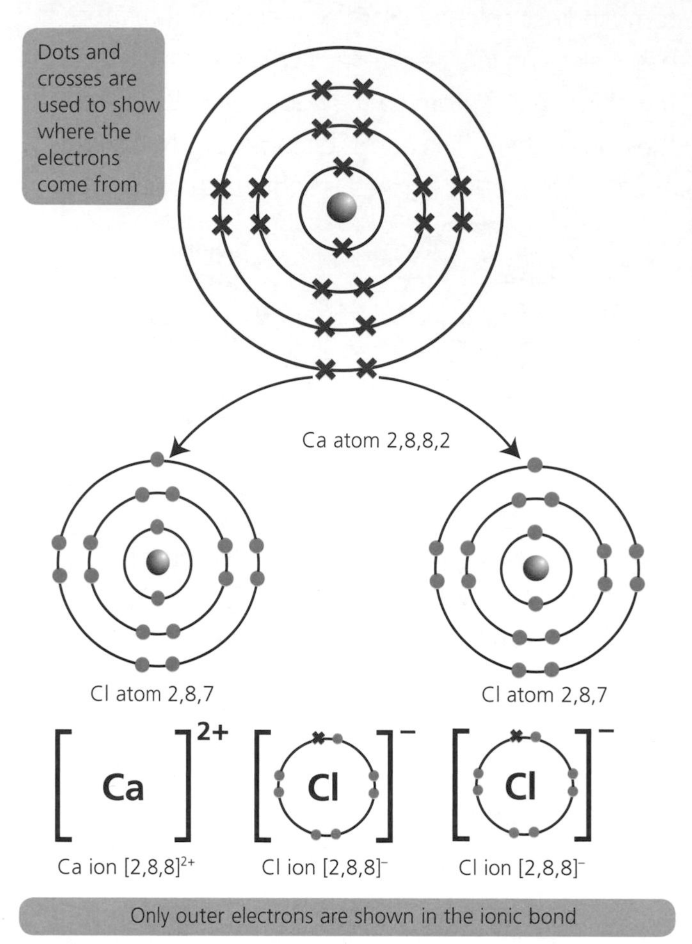

Only outer electrons are shown in the ionic bond

Example 3 – Magnesium and oxygen bond ionically to form magnesium oxide, MgO. The magnesium (Mg) atom has 2 electrons in its outer shell, which are transferred to the oxygen (O) atom so they both have 8 electrons in their outer shell. The atoms become ions Mg^{2+} and O^{2-} and the compound formed is magnesium oxide, MgO.

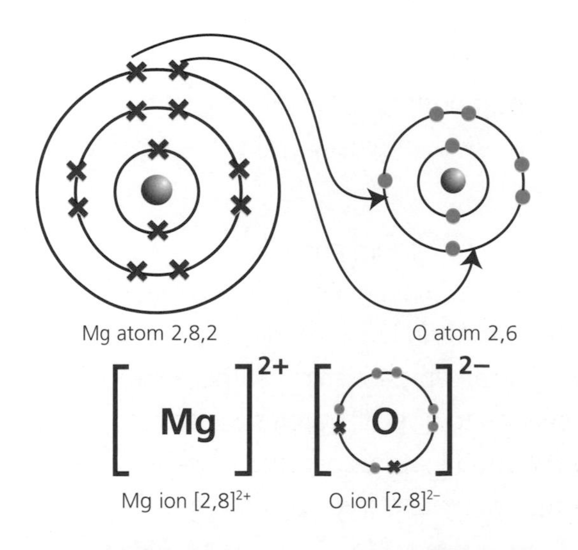

Covalent Bonds

A **covalent bond** is a very strong bond that is formed when electrons are **shared**. This occurs between non-metal atoms.

Some covalently bonded substances have simple molecules (like H_2, Cl_2, O_2, HCl, H_2O, CH_4) whereas others have giant covalent structures, called macromolecules (e.g. diamond, silicon dioxide).

A chlorine atom has 7 electrons in its outermost shell. In order to bond with another chlorine atom, an electron from each atom is shared to give both chlorine atoms 8 electrons in their outermost shell. This means each atom has a complete outer shell.

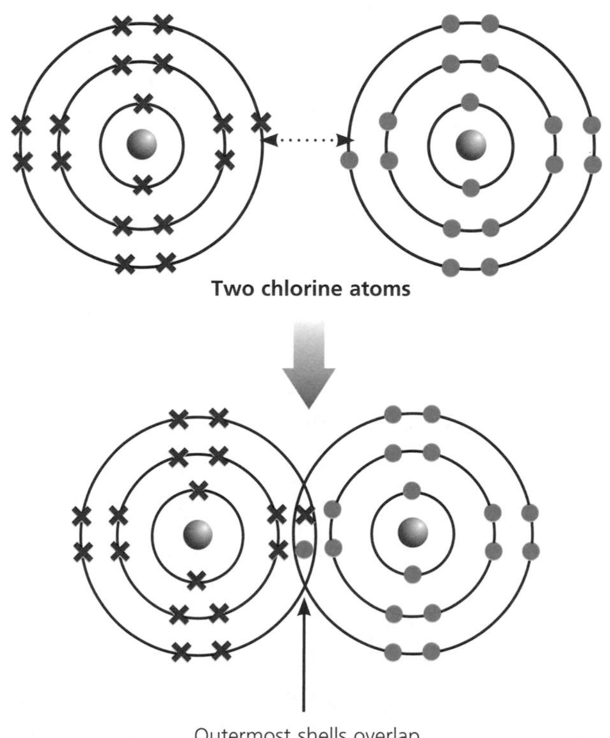

A chlorine molecule (Cl_2) (made up of two chlorine atoms)

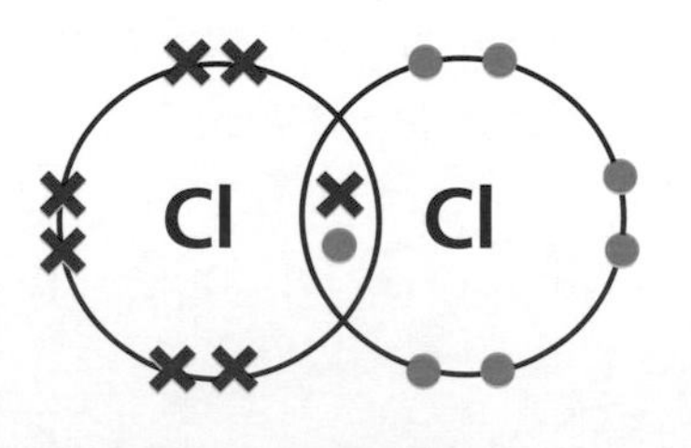

As with ionic bonds, only the outer shell of electrons need to be shown when drawing covalent bonds

Atoms that share electrons often form molecules in which there are strong covalent bonds between the atoms in each molecule but not between molecules. This means that these substances usually have low melting and boiling points.

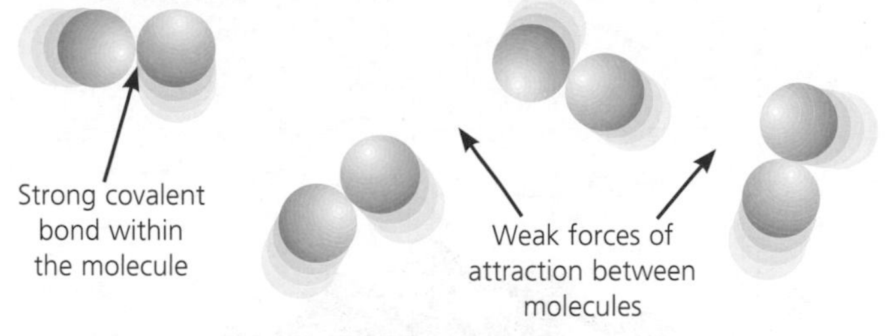

Covalent Bonding

You need to be familiar with the following examples, and know how to use three different methods for representing the covalent bonds in each molecule. Two forms are given in the examples below:

Water, H_2O		H–O–H
Chlorine, Cl_2		Cl – Cl
Hydrogen, H_2		H – H
Hydrogen chloride, HCl		H – Cl
Methane, CH_4		H–C–H with H above and below C
Oxygen, O_2		O=O (a double bond)

The third form of representing covalent bonds without using circles is shown here for an ammonia molecule (NH_3).

Simple Molecular Compounds

Gases, liquids and solids that have relatively low melting and boiling points consist of simple molecules. Because the molecules have no overall electric charge, they do not conduct electricity.

HT Simple molecular substances have low melting and boiling points because they have weak forces between molecules (intermolecular forces). It is these intermolecular forces that are broken when the substance melts or boils, rather than the strong covalent bonds holding the atoms together.

Giant Covalent Structures

As well as simple molecular substances, atoms that share electrons can also form giant structures (macromolecules). All the atoms in these giant structures are bonded to other atoms by strong covalent bonds, which means that all of these structures have very high melting and boiling points.

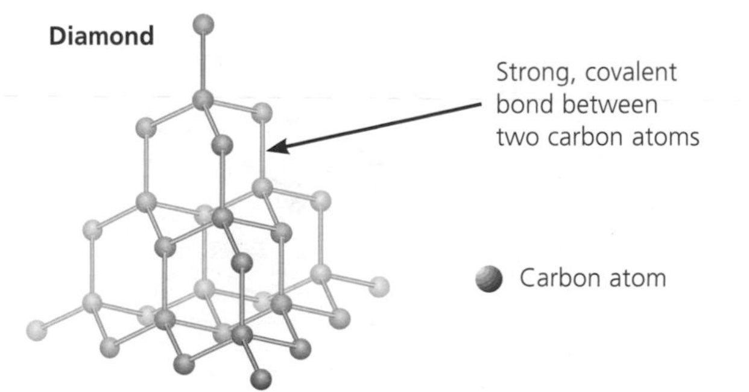

Diamond is a form of carbon that has a giant, rigid covalent structure (lattice) where each carbon atom forms four covalent bonds with other carbon atoms.

The large number of strong covalent bonds results in diamond having a very high melting point and which makes diamond very hard.

Giant Covalent Structures (cont)

Graphite

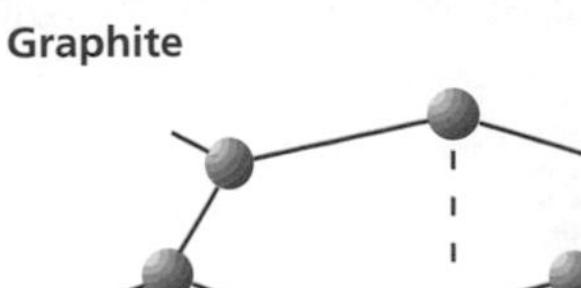

Graphite is a form of carbon that has a giant covalent structure (lattice). Unlike diamond, each carbon atom in graphite only forms three covalent bonds with other carbon atoms. This means that graphite has a layered structure with strong bonds between carbon atoms and weak intermolecular forces between layers. The layers can slide over each other because of the weak intermolecular forces, making graphite soft and slippery.

HT Because each carbon atom in graphite forms only three bonds, one electron from each atom is **delocalised**. These delocalised electrons are free to move and therefore graphite conducts electricity. It is the only non-metal element to do this.

Silicon Dioxide

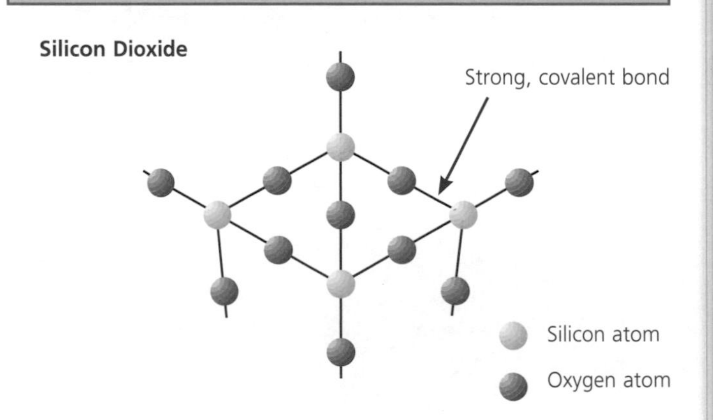

Silicon dioxide (SiO_2, also known as silica) has a giant, rigid covalent structure (lattice) similar to diamond, where each oxygen atom is joined to two silicon atoms and each silicon atom is joined to four oxygen atoms.

The large number of strong covalent bonds results in silicon dioxide having a very high melting point.

HT Carbon atoms also form fullerenes. These are hexagonal rings consisting of different numbers of carbon atoms joined together by covalent bonds. Fullerenes are not giant covalent structures, they are made up of simple molecules.

Fullerenes have many uses, for example:

- delivering drugs into the body
- nanotubes for reinforcing materials, e.g. tennis racquets.

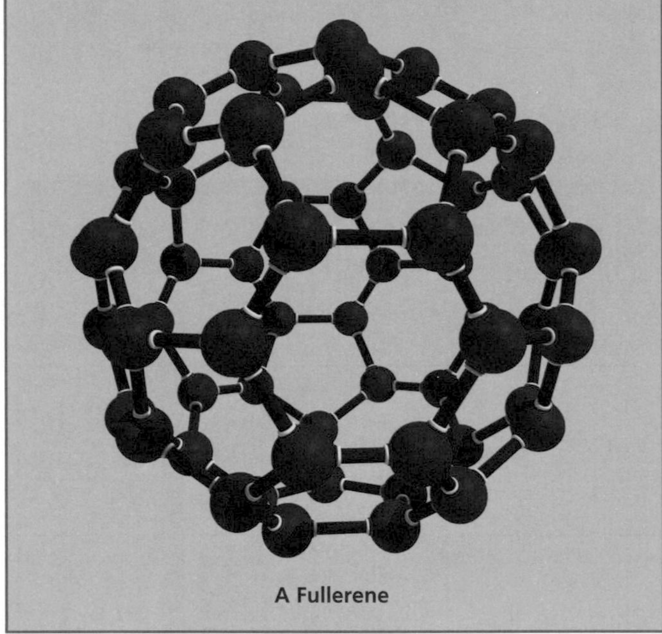

A Fullerene

Giant Ionic Structures

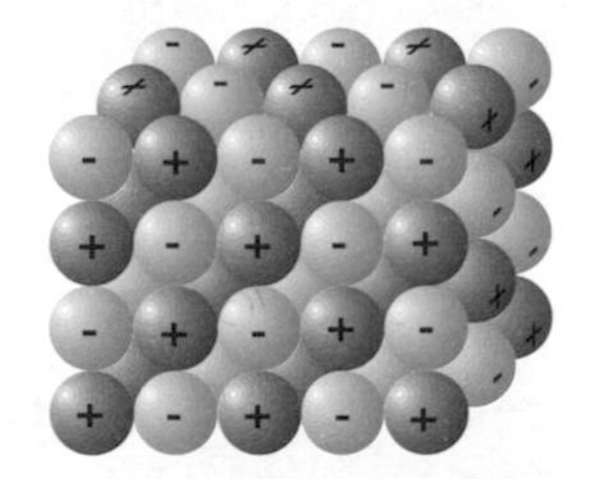

A giant ionic structure is a regular structure (giant ionic lattice) held together by the strong forces of attraction (electrostatic forces) between oppositely charged ions. These forces act in all directions in the lattice. This results in them having high melting and boiling points because large amounts of energy are required to break the strong ionic bonds.

Ionic compounds conduct electricity when molten or in solution because the charged ions are free to move about.

Metals

HT

Metals have a giant structure in which electrons in the highest energy level can be delocalised.

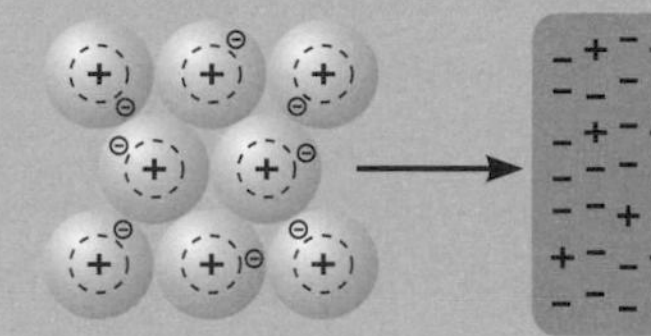

This effectively produces a regular arrangement (lattice) of positive ions held together by the electrostatic attraction between them and the delocalised electrons.

These delocalised electrons:

- hold the atoms together in a regular structure
- allow the atoms to slide over each other so metals can be bent and shaped
- can move around freely, which is why metals conduct heat and electricity.

Metals have giant structures made from atoms in a regular pattern. Alloys can be made from two or more different metals. Alloys are harder than pure metals because they contain different sized atoms, which distort the layers of atoms making it more difficult for them to slide over one another. Pure metals are more easily bent because atoms can slide over each other.

Shape-memory alloys have been developed to be able to return to their original shape after they have been deformed. Some examples of their uses are dental braces (Nitinol) and in glasses frames (see page 23).

Nanoparticles and Nanostructures

Nanoscience is the study of structures that are 1–100 nanometres in size, roughly in the order of a few hundred atoms. One nanometre is 0.000 000 001m (one billionth of a metre) and is written as 1nm or 1×10^{-9}m. (A human hair is around 20 000nm in width and a microorganism is around 200nm in diameter.)

Nanoparticles are tiny, tiny particles that can combine to form structures called **nanostructures**.

Nanostructures such as liposomes have always existed in naturally occurring substances. However, the technology to enable them to be seen did not exist until the early 1980s. Scientists working with new technology (the scanning tunnelling microscope) were able to construct enlarged images of surfaces, allowing them to see atoms and molecules for the first time. By the early 1990s, atoms could be isolated and moved. This means that nanostructures can be manipulated enabling materials to be developed that have new and specific properties that can be used in industry.

The properties of nanoparticles are different from the properties of the same materials in bulk. For example:

- electrons can move through an insulating layer of atoms
- nanoparticles are more sensitive to light, heat and magnetism
- nanoparticles possess a high surface area in relation to their volume.

A Magnified Representation of Iron Atoms in a Ring Around Some Surface State Electrons

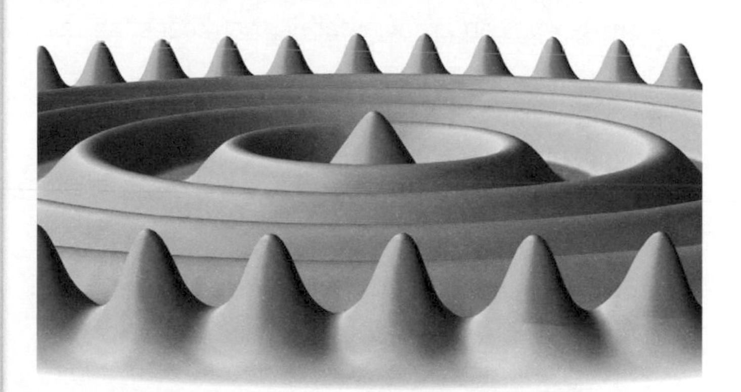

Nanocomposites

Other materials can be added to plastics to make stronger, stiffer and lighter materials called **nanocomposite** materials. The characteristics of nanocomposites can be seen by looking at the nanostructures formed by the nanoparticles.

Nanocomposites (cont)

Nanocomposites are already being used in the car industry and others are being developed with medical and dental applications in mind. They are also used in energy storage and separation processing, highly selective sensors, new coatings, sunscreens, drug delivery systems, stronger and lighter construction materials, textile fibres and product-specific catalysts.

Smart materials are a type of nanostructure that can be designed so they have specific properties on the nanoscopic scale or behave in a certain way when subjected to specific conditions.

Examples of Smart Materials

There are many different types of smart materials. Some examples are outlined below:

- **Lycra®** is a stretchy material used to make clothing. Lycra molecules have sections that make it both stretchy and rigid. It has properties that stop it being damaged by sweat and detergents; because it stretches to fit and then shrinks back on washing it makes it ideal for sport-type clothing.
- **Kevlar®** is an artificial (man-made) fibre used to make body armour. It is flexible, strong and tough. Its molecules pack closely together and bonds form between adjacent molecules, which make it strong.
- **Gore-Tex®** is an artificial fabric that keeps the rain out but lets sweat pass through. This is because it contains lots of tiny pores that stop water droplets from passing through the garment but allow water molecules from sweat to pass out. This means that it is a waterproof, breathable fabric, so is ideal for outdoor and sports clothing.

Polymers

Polymers are useful substances that have different properties depending on the raw materials used to make them and the conditions under which they were made. Poly(ethene) can be made using different catalysts to form both high density and low density forms.

Thermosetting and thermosoftening polymers differ in their structures. Thermosetting polymers consist of polymer chains held together by cross-links that make them strong. They do not change shape on heating and are used when plastics need to retain their shape when heated, e.g. plugs, hairdryers. Thermosoftening polymers do not have cross-links between the polymer chains and therefore their shape can be changed on heating, e.g. plastic bags.

Thermosetting

You need to be able to relate the properties of substances to their uses, and to suggest the type of structure of a substance when you are given its properties.

Substance	Properties	Example – Uses	Structure
Metal	• Strong. • Shiny. • Malleable (bendy). HT • Good conductor of heat and electricity.	• Steel – construction. • Gold – jewellery. • Aluminium – wires, aircraft. • Copper – pipes, wires.	• The layers of atoms can slide over each other. HT • Giant structure of atoms held together by metallic bonds allowing the outer electrons of each atom to move freely.
Non-metal	• Brittle. • Insulator.	• Glass – bottles. • Wood – pan handles.	• Covalent bonds.
Polymer	• Lightweight. • Flexible. • Waterproof.	• Poly(ethene) – plastic bags. • Shaped containers. • Polyvinyl chloride – rainwear.	• Long-chain structure of covalent bonded atoms. Forces between molecules are weak.
Ionic compound	• Hard, crystalline, often soluble in water. • High melting points. • Insulator when solid but conducts electricity when molten or dissolved.	• Sodium chloride – food additive. • Sodium chloride – electrolysis to produce chlorine and sodium hydroxide.	• Force of attraction between oppositely charged ions formed by electron transfer. A lattice results that is difficult to break down. In the solid state the ions are held in place. Once melted or dissolved they are free to move.
Molecular covalent	• Soft. • Low melting points. • Insulator	• Ammonia, nitrogen and oxygen gas, e.g. ammonia is used to make fertiliser.	• Molecules with no charge. • Strong bonding inside each molecule but bonding between molecules is weak.
Macromolecule	• Hard. • High melting points.	• Diamond – drill heads. • Silicon dioxide – sand for mortar and concrete. HT • Graphite, fullerenes.	• Giant covalent molecule. • Huge lattices with millions of covalent bonds. • No weak forces to break.
Nanomaterial	• Specific to use.	• Nanoparticles – catalysts. • Nanotubes – reinforce tennis racquets, used in computer chips.	• Really tiny particles shaped like hollow balls or closed tubes. • Atoms form covalent bonds leaving free electrons.
Smart material	• Specific to use.	• Nitinol – spectacle frames.	• Can exist in two different solid forms. The molecules absorb energy to rearrange the atoms into a new form.

You need to be able to evaluate developments and applications of new materials, e.g. nanomaterials, smart materials.

New materials are being developed to provide us with materials that have advantageous properties and can therefore be very useful.

The development of nanomaterials is part of nanotechnology (the understanding and control of very small matter). Nanomaterials have many properties, which means they have many uses in industry, for example as industrial catalysts. Their very, very small size means they have a large surface area in relation to their size.

Individual nanoparticles have different properties from the whole chemical. They can:

- fill plastics and coat surfaces
- absorb and reflect harmful ultraviolet rays in suncreams and cosmetics
- be added to glass to repel water, to keep windows cleaner for longer
- be released in washing machines to clean clothes thoroughly
- be released in fridges to kill microorganisms and keep food fresher for longer
- be used in sensors, e.g. to test water purity.

Nanotubes which join nanoparticles are very strong. They can conduct electricity so they can be used in electric circuits.

Nanotechnology has a wide variety of potential applications in biomedical, optical and electronic fields.

For example, nanotechnology could be used to create secure communication systems, detect and eradicate small tumours, help in the diagnosis of diseases, and in the development of microscopic surgery that would not leave scars.

Nanomaterials can be developed so they have useful properties such as being able to change shape or size as a result of being heated, or changing from a liquid to a solid when near a magnet. These are a type of smart material and they can be categorised into different groups – electroactive, thermoactive, and magnetoactive – depending on the trigger they respond to. Each type has a different property that can be altered and therefore they each have a different application.

Smart materials can be used in sportswear because they are good thermal insulators as well as being lightweight, breathable and waterproof.

New Material	Advantages	Disadvantages
Smart materials	• Many different properties. • Many applications. • Easy to manipulate. • Many potential uses to investigate.	• Cost of developing new materials.
Nanomaterials	• Many applications. • Can reduce costs, e.g. when used as catalysts. • Have the potential for many more beneficial uses, especially in the medical industry.	• Difficult to engineer nanoparticles. • Can be dangerous in certain situations, e.g. if they get into drinking water.

Atomic Structure, Analysis and Quantitative Chemistry C2

> **C2.3 Atomic structure, analysis and quantitative chemistry**
>
> Atomic masses can be used to calculate the yield from a chemical reaction because we know that no atoms are lost or gained. We can use instrumental techniques to separate and identify compounds and elements in a mixture. To understand this, you need to know:
>
> - how to find the mass number and atomic number of an element
> - how to calculate relative atomic mass and relative formula mass
> - how to use the relative atomic mass and relative formula mass to find the percentage of an element in a compound
> - that some reactions can be reversible
> - that gas chromatography and mass spectroscopy can be used to identify and separate compounds from a mixture.

Mass Number and Atomic Number

Atoms of an element can be represented very conveniently. For example, take the sodium atom:

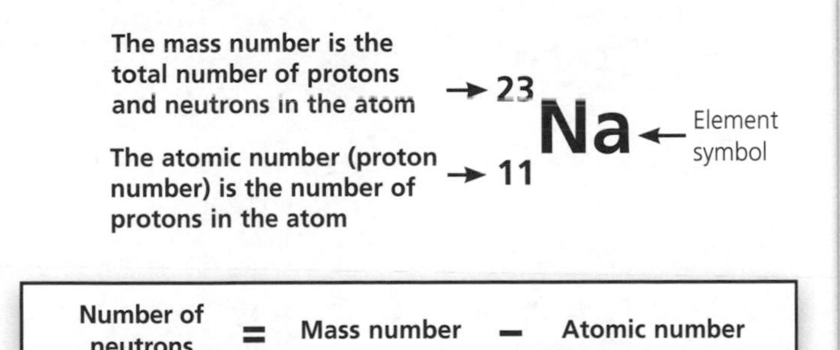

Number of neutrons	=	Mass number	–	Atomic number

The **atomic number** gives the number of protons, which is equal to the number of electrons. Because atoms have the same number of protons and electrons, they have no overall charge.

Examples

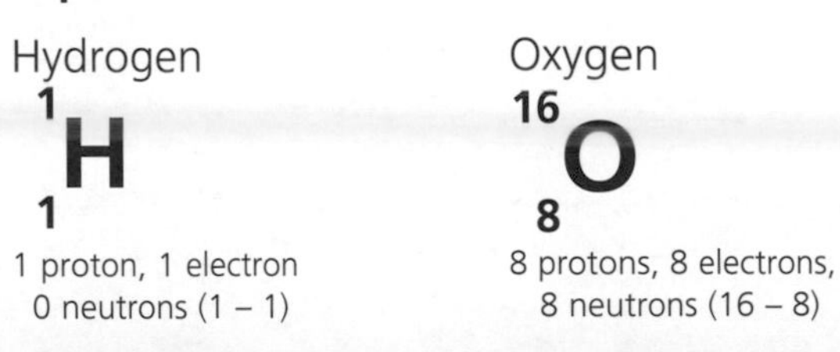

Although protons and electrons balance each other out because they have opposite charges, they do not have equal mass. Protons and neutrons (which together form the nucleus) each have a **relative mass** of 1, whereas the relative mass of an electron is almost nothing.

Atomic Particle	Relative Mass
Proton	1
Neutron	1
Electron	Very small (negligible)

Isotopes

All atoms of a particular element have the same number of protons; atoms of different elements have different numbers of protons.

However, some atoms of the same element can have different numbers of neutrons. These are called **isotopes**. Isotopes have the same atomic number but different mass numbers.

Examples

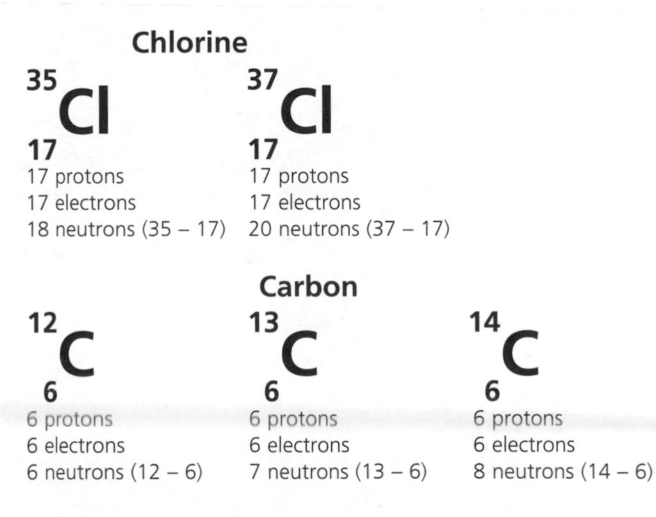

Relative Atomic Mass, A_r

Atoms are too small for their actual atomic mass to be of much use to us. To make things more manageable we use **relative atomic mass, A_r**.

HT The relative atomic mass is the mass of a particular atom compared to a twelfth of the mass of a carbon atom (the ^{12}C isotope). It is an average value for all the isotopes of the element.

Relative Formula Mass, M_r

The **relative formula mass (M_r)** of a compound is simply the relative atomic masses of all its elements added together. To calculate M_r, we need the formula of the compound and the A_r of all the atoms involved.

Example 1
Using the data below, calculate the M_r of water, H_2O.

Because water has an M_r of 18, it is 18 times heavier than a hydrogen atom, or one and a half times heavier than a carbon atom, or two-thirds as heavy as an aluminium atom.

Example 2
Using the data below, calculate the M_r of potassium carbonate, K_2CO_3.

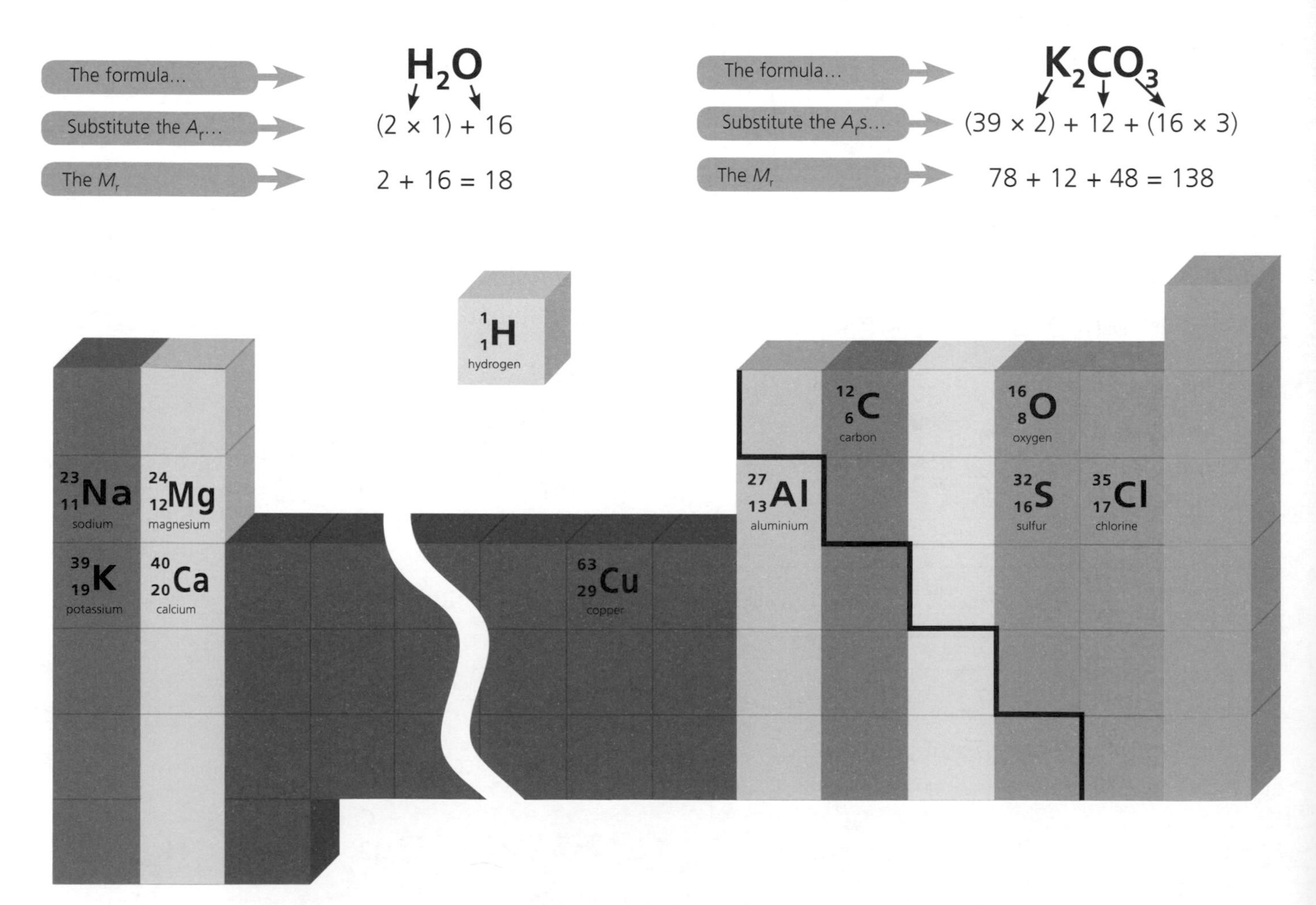

Calculating Percentage Mass of an Element in a Compound

If there are 12 left-handed pupils in a class of 30, you can work out the percentage of left-handers in the following way:

$$\frac{\text{Number of left-handers}}{\text{Total number in class}} \times 100\%$$

In this case:

$\frac{12}{30} \times 100\% = \mathbf{40\%}$

You use exactly the same principle to calculate the percentage mass of an element in a compound, except this time we express it as:

$$\frac{\textbf{Relative mass of element in the compound}}{\textbf{Relative formula mass of compound } (M_r)} \times 100\%$$

All you need to know is the formula of the compound and the relative atomic masses of all the atoms.

Example 1

Calculate the percentage mass of magnesium in magnesium oxide, MgO.

Relative mass of magnesium = 24

Relative formula mass (M_r) of MgO =

$$24 + 16 = 40$$

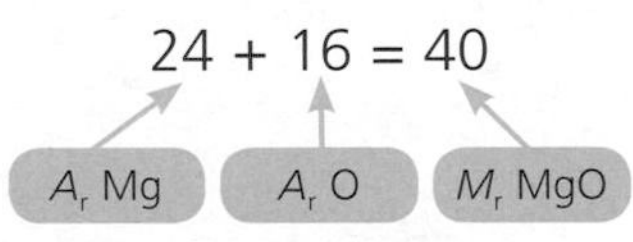

Substituting into our formula:

$\frac{\text{Relative mass of element}}{M_r \text{ of compound}} \times 100\%$

$\frac{24}{40} \times 100\% = \mathbf{60\%}$

Example 2

Calculate the percentage mass of potassium in potassium carbonate, K_2CO_3.

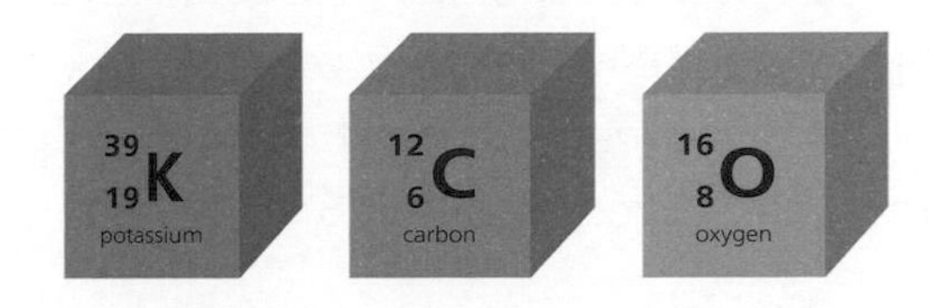

Relative mass of potassium = 39 × 2 = 78

Relative formula mass (M_r) of K_2CO_3 =

$$78 + 12 + 48 = 138$$

A_r K × 2 | A_r C | A_r O × 3 | M_r K_2CO_3

Substituting into our formula:

$\frac{\text{Relative mass of element}}{M_r \text{ of compound}} \times 100\%$

$\frac{78}{138} \times 100\% = \mathbf{56.5\%}$

HT

Calculating the Empirical Formula of a Compound

The empirical formula of a compound is the simplest whole number formula that represents the composition of the compound by mass.

Example

Find the empirical formula of an oxide of iron produced by reacting 1.12g of iron with 0.48g of oxygen (A_r Fe = 56, A_r O = 16).

Identify the actual masses of the elements in the compound:

Masses: Fe = 1.12, O = 0.48

Divide these masses by their relative atomic masses:

$\text{Fe} = \frac{1.12}{56} = 0.02$ $\text{O} = \frac{0.48}{16} = 0.03$

Divide each of these by the smallest to get the ratio:

$\text{Fe} = \frac{0.02}{0.02} = 1$ $\text{O} = \frac{0.03}{0.02} = 1.5$

If the ratio does not work out as whole numbers then multiply each number by the same factor to get whole numbers:

Fe = 1 × 2 = 2 O = 1.5 × 2 = 3

Empirical formula = $\mathbf{Fe_2O_3}$

The Mole

A **mole** (**mol**) is a measure of the number of particles (atoms or molecules) contained in a substance. One mole of any substance (element or compound) will always contain the same number of particles – six hundred thousand billion billion or 6×10^{23}.

The mass of one mole of an element is its relative atomic mass, A_r, or relative formula mass (M_r) in grams.

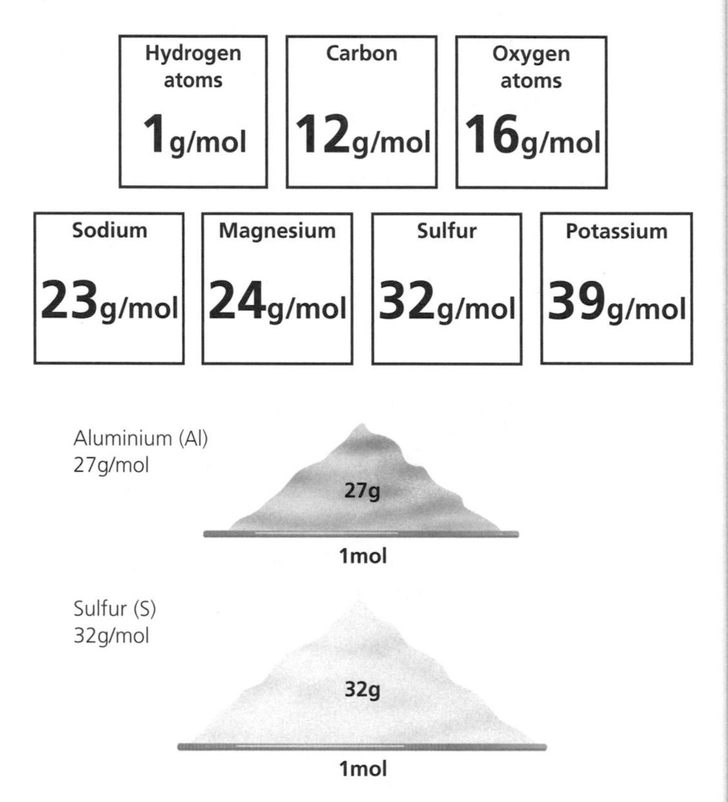

If a substance is a compound, the mass of one mole of the substance is always equal to the relative formula mass, M_r (A_rs of all its elements added together), of the substance in grams.

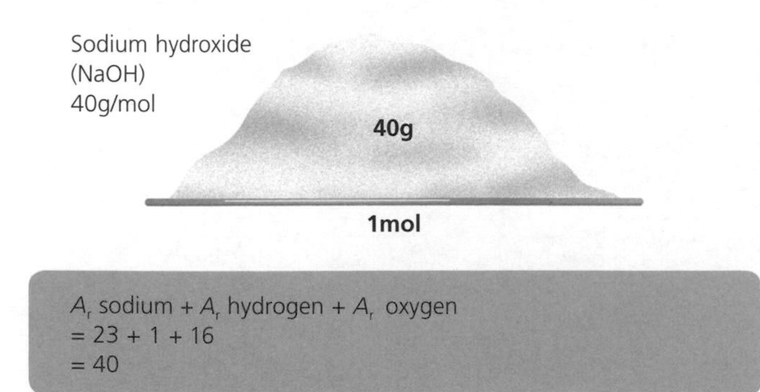

A_r sodium + A_r hydrogen + A_r oxygen
= 23 + 1 + 16
= 40

HT Questions involving moles can be answered using the following relationship. You need to remember this because it will not be given to you in the examination.

$$\text{Moles} = \frac{\text{Mass}}{M_r}$$

Example 1

Calculate the number of moles of carbon (C) in 36g of the element.

Using the relationship:

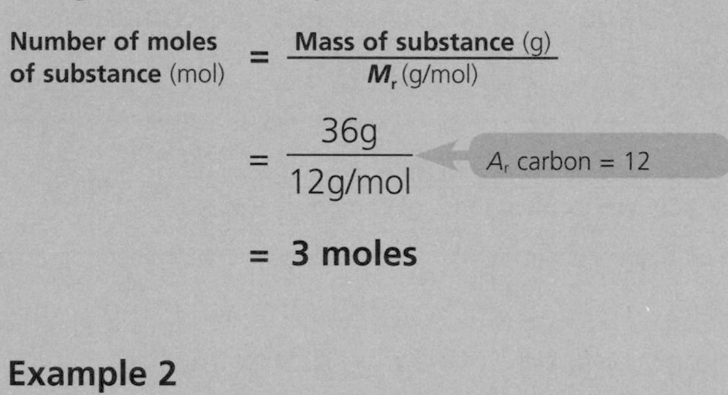

$$\text{Number of moles of substance (mol)} = \frac{\text{Mass of substance (g)}}{M_r\text{ (g/mol)}}$$

$$= \frac{36\text{g}}{12\text{g/mol}}$$

A_r carbon = 12

$$= \textbf{3 moles}$$

Example 2

Calculate the number of moles of carbon dioxide (CO_2) in 33g of the gas.

Using the relationship:

$$\text{Number of moles of substance (mol)} = \frac{\text{Mass of substance (g)}}{M_r\text{ (g/mol)}}$$

$$= \frac{33\text{g}}{44\text{g/mol}}$$

$$= \textbf{0.75 mole}$$

M_r carbon dioxide
= A_r carbon +
2 × A_r oxygen
= 12 + (2 × 16)
= 44

Example 3

Calculate the mass of 4 moles of sodium hydroxide (NaOH).

Rearranging the relationship:

$$\text{Mass of substance (g)} = \text{Number of moles of substance (mol)} \times M_r\text{ (g/mol)}$$

$$= 4\text{mol} \times 40\text{g/mol}$$

$$= \textbf{160g}$$

Calculating the Mass of a Product

Example

Calculate how much calcium oxide can be produced from 50kg of calcium carbonate. (Relative atomic masses: Ca = 40, C = 12, O = 16).

Write down the equation:

$$CaCO_3 \rightarrow CaO + CO_2$$

Work out the M_r of each substance:

$$40 + 12 + (3 \times 16) \rightarrow (40 + 16) + [12 + (2 \times 16)]$$

Check the total mass of the reactants equals the total mass of the products. If they are not the same, check your work:

$$100 \rightarrow 56 + 44 \checkmark$$

Since the question only mentions calcium oxide and calcium carbonate, you can now ignore the carbon dioxide. You just need the ratio of mass of reactant to mass of product.

100 : 56

If 100kg of $CaCO_3$ produces 56kg of CaO, then 1kg of $CaCO_3$ produces $\frac{56}{100}$kg of CaO, and 50kg of $CaCO_3$ produces $\frac{56}{100} \times 50 =$ **28kg** of CaO.

Calculating the Mass of a Reactant

Example

Calculate how much aluminium oxide is needed to produce 540 tonnes of aluminium. (Relative atomic masses: Al = 27, O = 16).

Write down the equation:

$$2Al_2O_3 \rightarrow 4Al + 3O_2$$

Work out the M_r of each substance:

$$2[(2 \times 27) + (3 \times 16)] \rightarrow (4 \times 27) + [3 \times (2 \times 16)]$$

Check the total mass of reactants equals the total mass of the products:

$$204 \rightarrow 108 + 96 \checkmark$$

Since the question only mentions aluminium oxide and aluminium, you can now ignore the oxygen. You just need the ratio of mass of reactant to mass of product.

204 : 108

If 204 tonnes of Al_2O_3 produces 108 tonnes of Al, then $\frac{204}{108}$ tonnes is needed to produce 1 tonne of Al, and $\frac{204}{108} \times 540$ tonnes is needed to produce 540 tonnes of Al, i.e. **1020 tonnes** of Al_2O_3 is needed.

Yield

Atoms are never lost or gained in a chemical reaction. However, it is not always possible to obtain the calculated amount of a product because:

- if the reaction is reversible, it may not go to completion
- some product could be lost when it is separated from the reaction mixture
- there could be different ways for the reactants to behave in an expected reaction.

The amount of product actually obtained is called the **yield**. The **percentage yield** can be calculated by comparing the actual yield obtained from a reaction with the maximum theoretical yield.

$$\text{Percentage yield} = \frac{\text{Yield from reaction}}{\text{Maximum theoretical yield}} \times 100$$

Example

We know from the example above that you would expect to produce 28kg of calcium oxide (CaO) from 50kg of calcium carbonate ($CaCO_3$). This is the maximum theoretical yield.

A company heats 50kg of calcium carbonate in a kiln and obtains 22kg of calcium oxide. What is the percentage yield?

Using the formula, the percentage yield is:

$$\text{Percentage yield} = \frac{22}{28} \times 100$$

$$= \mathbf{78.6\%}$$

C2 Atomic Structure, Analysis and Quantitative Chemistry

Reversible Reactions

Some chemical reactions are **reversible**, i.e. the products can react to produce the original reactants.

A and B react to produce C and D, but C and D can also react to produce A and B. For example:

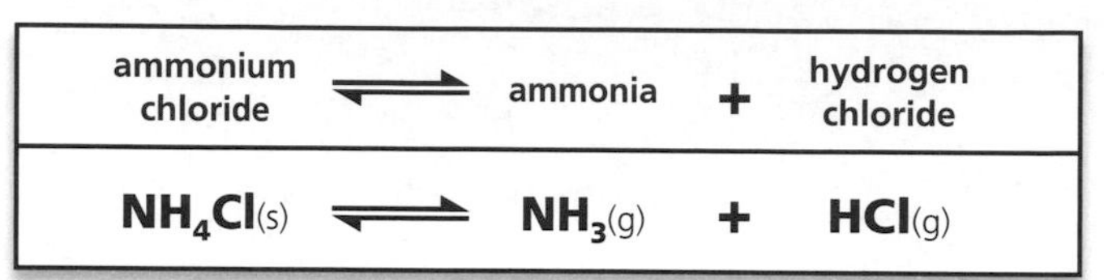

ammonium chloride	⇌	ammonia	+	hydrogen chloride
$NH_4Cl(s)$	⇌	$NH_3(g)$	+	$HCl(g)$

Solid ammonium chloride decomposes when heated to produce ammonia and hydrogen chloride gas, both of which are colourless.

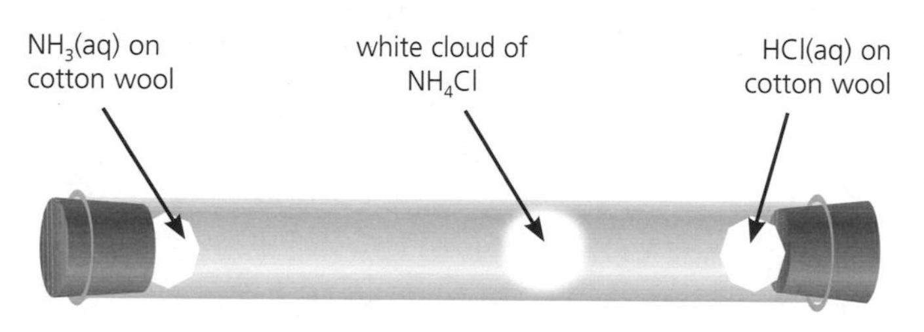

Ammonia reacts with hydrogen chloride gas to produce clouds of white ammonium chloride powder.

Analysing Substances

Instrumental methods such as chromatography and mass spectrometry can be used to detect and identify elements and compounds. They are accurate, sensitive, quick and allow very small samples to be used.

Chromatography

Chemical analysis is used to separate and identify additives in food – for example, chromatography is used to identify artificial colours.

Chromatography allows us to identify unknown substances by comparing them to known substances.

Example

Samples of four known substances (A, B, C and D) and an unknown substance (X) are placed on a pencil line on a piece of chromatography paper. The paper is then stood in a solvent (below the pencil line). As the solvent travels up the paper, it dissolves the samples and carries them up the paper.

The substances will move up the paper by differing amounts due to their different solubilities in the solvent. Substance X is identified by comparing its 'spots' with those of A–D.

Apparatus for Chromatography

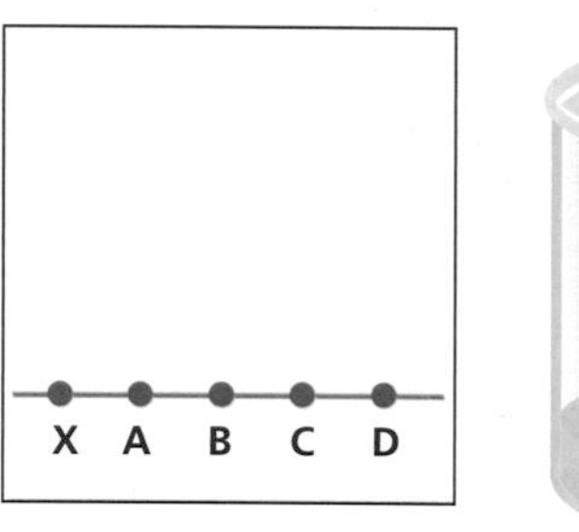

Spots of food colouring on filter paper

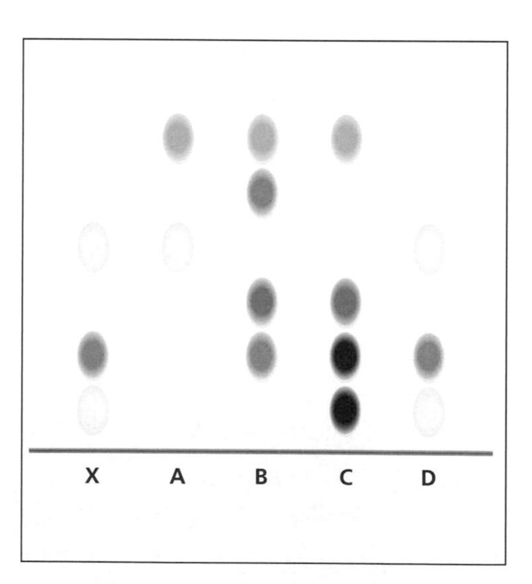

A Chromatogram Showing how the Food Colourings have Split into their Separate Dyes

By comparing food colourings A, B, C and D to substance X, we can see that substance X is food colouring D.

Gas Chromatography

Gas chromatography is used to separate and identify mixtures. A mixture of compounds is carried by a gas through a column packed with powdered material. The compounds in the mixture will travel at different speeds and this causes them to separate.

A chromatogram (the display printed out by the gas chromatography machine) shows the number of compounds present by the number of peaks. The retention time of each compound is indicated by the position of each peak.

The retention time is the time taken for a substance to travel through the column. This can be used to help identify the substance injected.

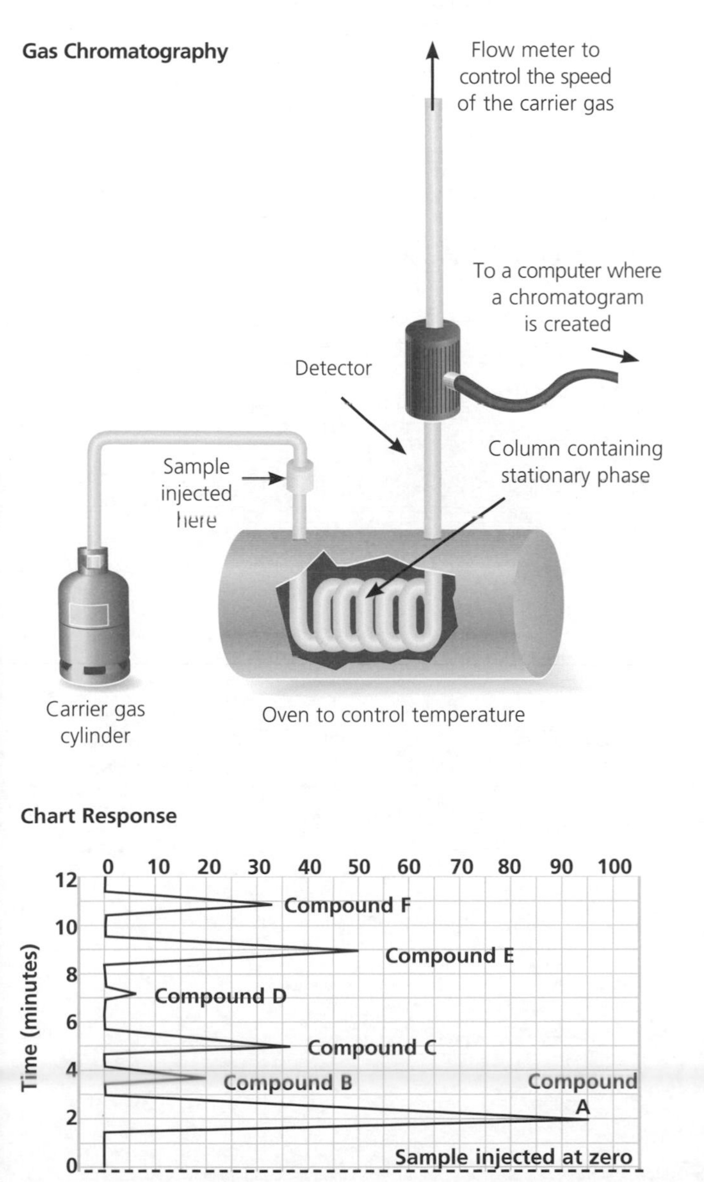

The gas chromatogram shows that there are six different compounds present in the sample injected. The size of each peak shows the relative amount of each compound (A is present in the largest amount and D in the smallest amount). The distance of the peak (time) from the injection time gives the retention time for each compound: A has the shortest retention time, F the longest. This means that it has taken compound A the least time to pass through the column and compound F the most time.

Mass Spectrometry

To aid the identification of the sample injected, a mass spectrometer is often linked to the gas chromatography column. This allows the substances leaving the column to be identified.

Mass spectrometers identify substances very quickly and accurately even when very small quantities are used. The mass spectrometer can give the relative formula mass (M_r) of each of the substances separated by gas chromatography if the two techniques are used together.

HT Mass spectrometry produces a 'graph' for each compound showing peaks. The relative formula mass can be identified from the molecular ion peak, usually the peak to the far right of the graph.

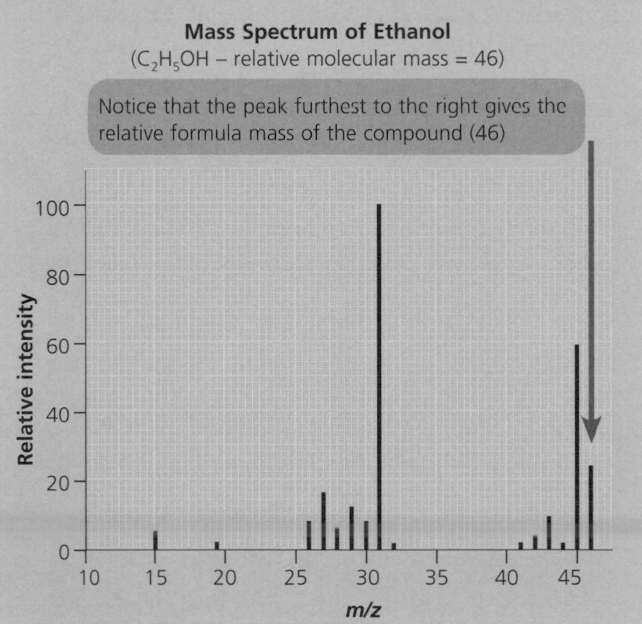

You need to be able to calculate chemical quantities involving relative formula mass (M_r) for industrial processes and evaluate sustainable development issues relating to this economy.

Ethyl ethanoate is used as a solvent in glues and nail polish removers. It is produced in the reaction:

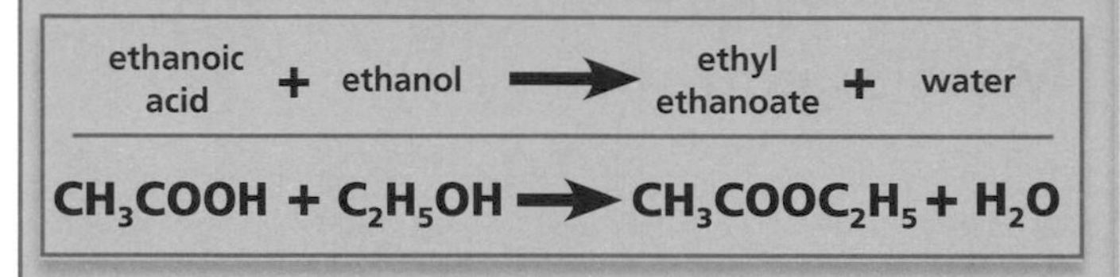

ethanoic acid + ethanol → ethyl ethanoate + water

$$CH_3COOH + C_2H_5OH \rightarrow CH_3COOC_2H_5 + H_2O$$

In industry, research scientists measure or calculate the amounts of materials used and produced in reactions to make sure that reactions are economical. If a reaction does not convert all the reactants into useful products, there will be some wastage. As a result, the manufacturer might have to raise the sell-on price of the product, and this will have an impact on the retail price of all the final products made using it.

It is important that chemical reactions that are part of industrial processes are as economical as possible, so that the product costs the people who need to buy it as little as possible.

It is also important to plan for meeting the present needs of people without spoiling the environment. This is called a sustainable development. The Earth only has a finite supply of minerals, so it makes sense to only use as little as we need. Therefore, industries aim to use smaller amounts of raw materials and energy while creating less waste.

Scientists in industry look for ways to use the waste products so that they become useful by-products of the reactions.

The chemical industry examines its processes carefully to make sure that it:

- makes efficient use of energy
- reduces the hazards and risks of the chemicals it uses and makes
- reduces waste
- attempts to convert a high proportion of the atoms in the reactants into the products
- uses mainly renewable resources
- prevents pollution of the environment.

Example – Calculating Percentage Yield

920g of ethanol is reacted with excess ethanoic acid to make ethyl ethanoate. 1320g of ethyl ethanoate are produced. Calculate the % yield for the reaction.

Write down the equation:

ethanoic acid + ethanol → ethyl ethanoate + water

$$CH_3COOH + C_2H_5OH \rightarrow CH_3COOC_2H_5 + H_2O$$

Work out the number of moles of ethanol used and therefore the number of moles of ethyl ethanoate that could be produced

Number of moles of ethanol (C_2H_5OH)

$$= \frac{\text{Mass of ethanol (g)}}{M_r \text{ of ethanol}} = \frac{920}{46} = 20 \text{ mol}$$

1 mole of ethanol produces 1 mole of ethyl ethanoate, therefore, 20 moles of ethyl ethanoate could be produced.

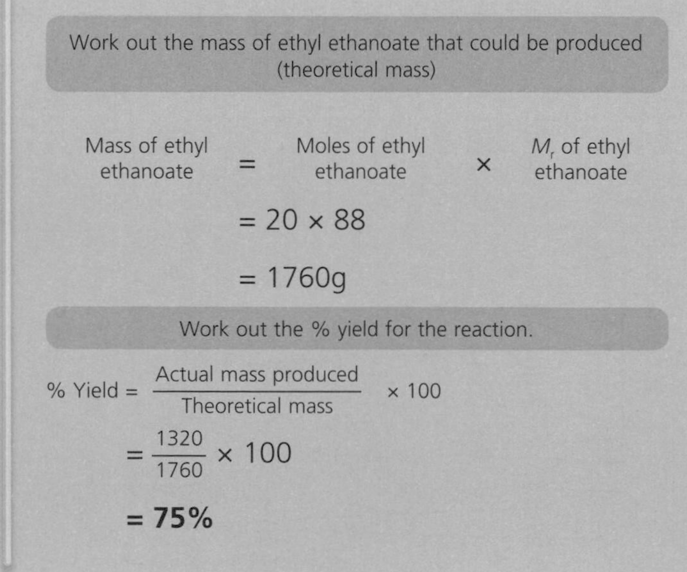

Work out the mass of ethyl ethanoate that could be produced (theoretical mass)

$$\text{Mass of ethyl ethanoate} = \text{Moles of ethyl ethanoate} \times M_r \text{ of ethyl ethanoate}$$

$$= 20 \times 88$$

$$= 1760\text{g}$$

Work out the % yield for the reaction.

$$\% \text{ Yield} = \frac{\text{Actual mass produced}}{\text{Theoretical mass}} \times 100$$

$$= \frac{1320}{1760} \times 100$$

$$\mathbf{= 75\%}$$

C2.4 Rates of reaction

Controlling the rates of chemical reactions is very important in both everyday life and industry. To understand this, you need to know:

- how to determine a rate of reaction
- what factors affect the rate of a reaction
- how and why catalysts are used to alter the rate of chemical reactions.

Rates of Reactions

Chemical reactions only occur when reacting particles collide with each other with sufficient energy. The minimum amount of energy required to cause a reaction is called the **activation energy**. There are five important factors that affect the rate of a reaction:

- temperature
- concentration of solutions
- pressure of gases
- surface area of solids
- use of a catalyst.

Temperature of the Reactants

Low Temperature	High Temperature
In a cold reaction mixture, the particles are moving relatively slowly – the particles will collide with each other less often, with less energy, and so few collisions will be successful. 	If we heat the reaction mixture, the particles will move more quickly – the particles will collide with each other more often, with greater energy, and so many more collisions will be successful. 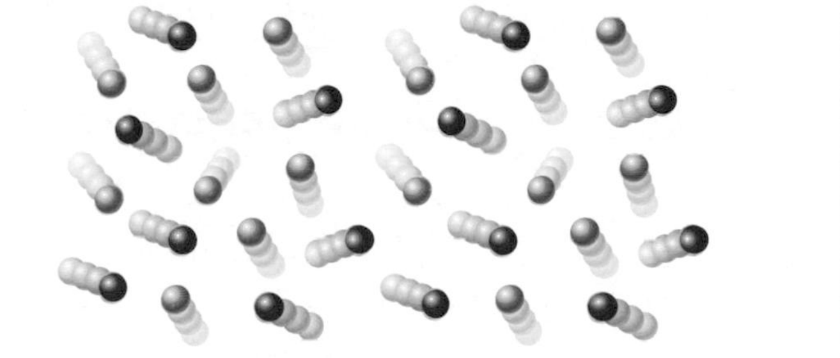

Concentration of Dissolved Reactants

Increasing the concentration of one or more of the reactants means there are more particles of that reactant in the same volume.

Low Concentration	High Concentration
In a reaction where one or both reactants are in low concentrations, the particles are spread out – the particles will collide with each other less often resulting in few successful collisions. 	Where there are high concentrations of one or both reactants, the particles are crowded close together – the particles will collide with each other more often, resulting in many more successful collisions. 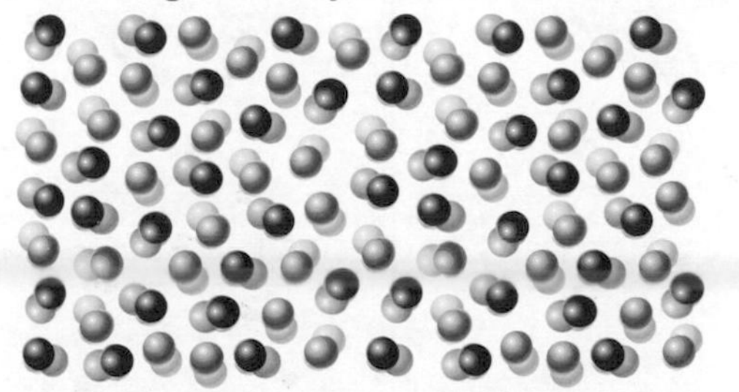

Concentration (cont)

We see a similar effect when the reactants are gases. As the pressure on a gas is increased, the particles are pushed closer together so they collide more often and the reaction is faster.

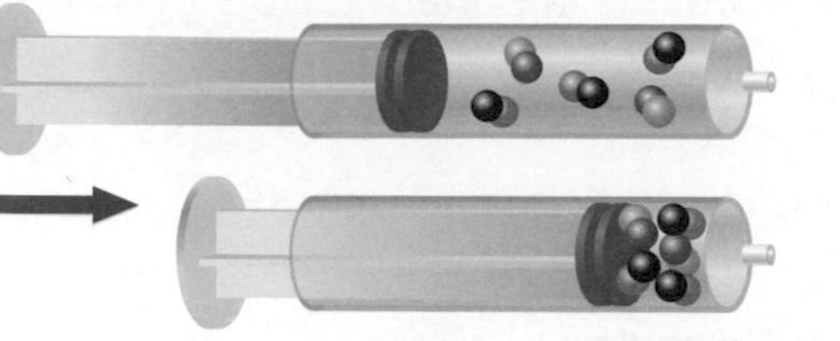

HT **Concentrations** of solutions are given in **moles per cubic decimetre (mol/dm³)**. Equal volumes of solutions with the same concentration contain the same number of moles of solute, i.e. the same number of particles.

Equal volumes of gases at the same temperature and pressure contain the same number of particles.

Surface Area of Solid Reactants

Large pieces of a solid have a small surface area in relation to their volume, so fewer particles are exposed and available for collisions. This means fewer collisions and a slower reaction. Small pieces of a solid have a large surface area in relation to their volume, so more particles are exposed and available for collisions. This means more collisions and a faster reaction.

Large Pieces	Small Pieces
• Small surface area. • Fewer collisions. • Reaction rate is slow.	• Large surface area. • More collisions. • Reaction rate is faster.

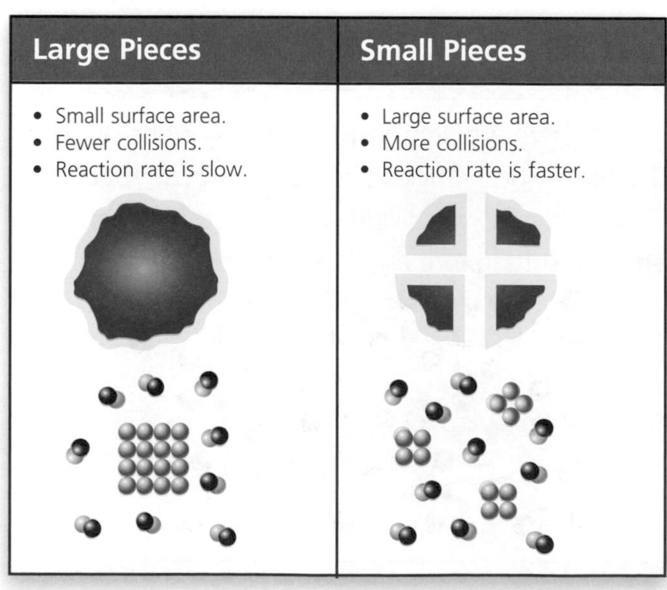

Using a Catalyst

A **catalyst** is a substance that increases the rate of a chemical reaction without being used up or altered in the process. It can be used over and over again to increase the rate at which reactants are converted into products.

A catalyst lowers the amount of energy needed for a successful collision, so more collisions will be successful and the reaction will be faster. One way it can do this is to provide a surface for the molecules to attach to, thereby increasing their chances of bumping into each other.

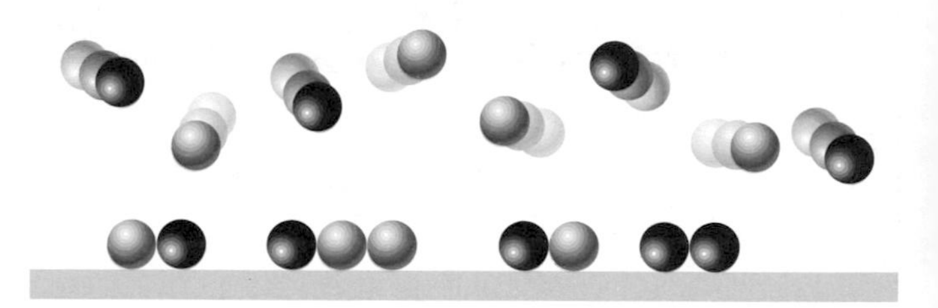

Catalysts are specific to certain reactions, i.e. different reactions need different catalysts – e.g. the cracking of hydrocarbons uses broken pottery; the manufacture of ammonia (Haber process) uses iron.

Increasing the rates of chemical reactions is important in industry because it helps to reduce costs.

Lowering activation energy means reactions can happen at a lower temperature, which also reduces costs.

Catalysts Used in Industrial Reactions

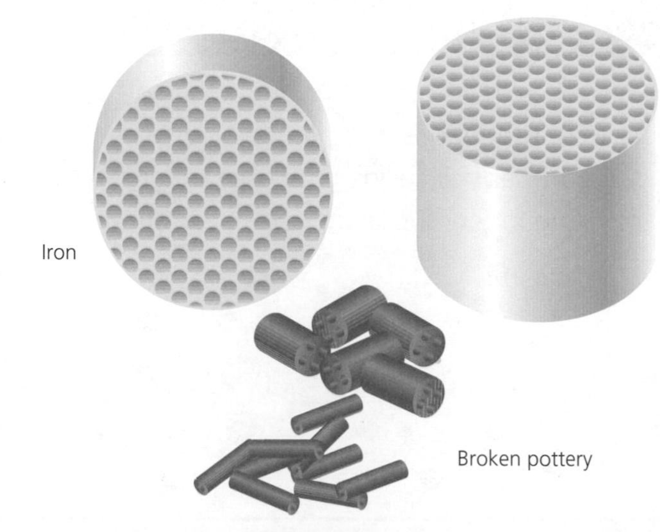

Analysing the Rate of Reaction

The rate of a chemical reaction can be found by:

- measuring the amount of reactants used or measuring the amount of products formed
- measuring the time taken.

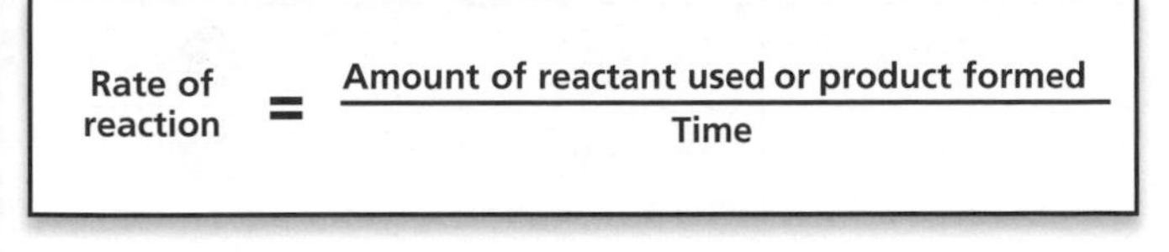

For example, the decomposition of hydrogen peroxide using manganese (IV) oxide.

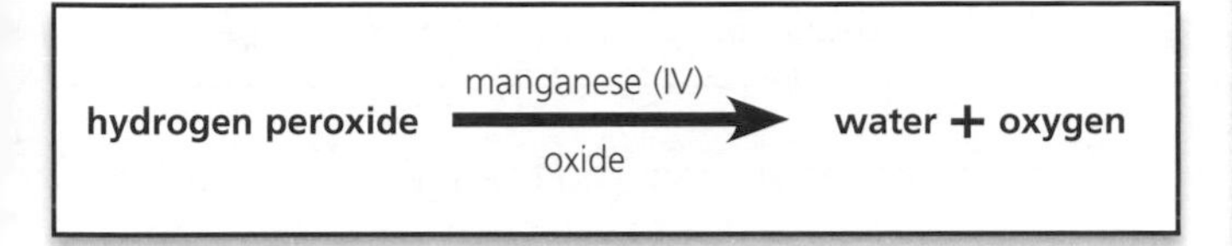

You could measure the amount of product formed by weighing the mixture before and after the reaction takes place. In the case of hydrogen peroxide, oxygen is released from the mixture so the mass of the mixture will decrease over time.

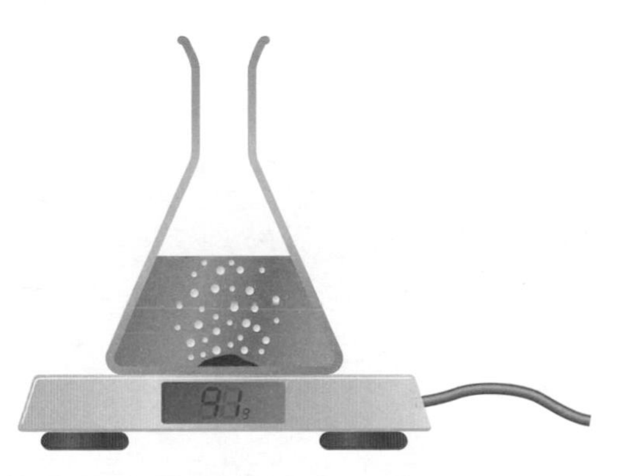

You could use a gas syringe to measure the volume of gas produced, in this case, oxygen.

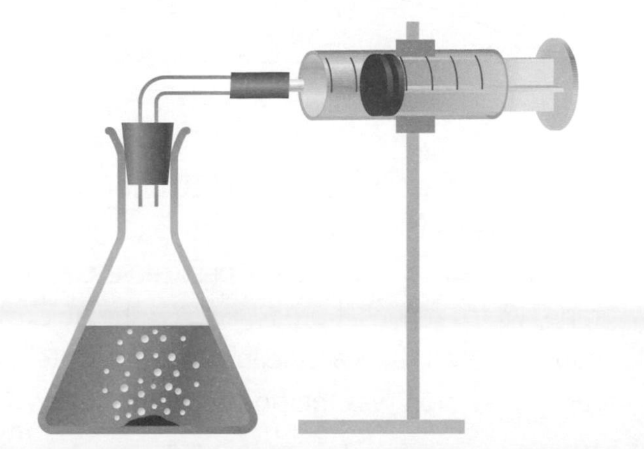

To test for oxygen gas, insert a glowing splint into a jar of collected gas. Oxygen will relight a glowing splint.

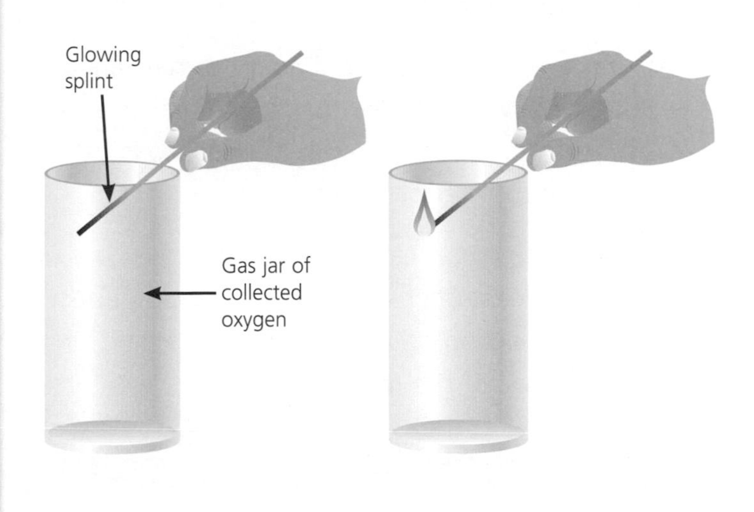

Graphs can then be plotted to show the progress of a chemical reaction – there are three things to remember:

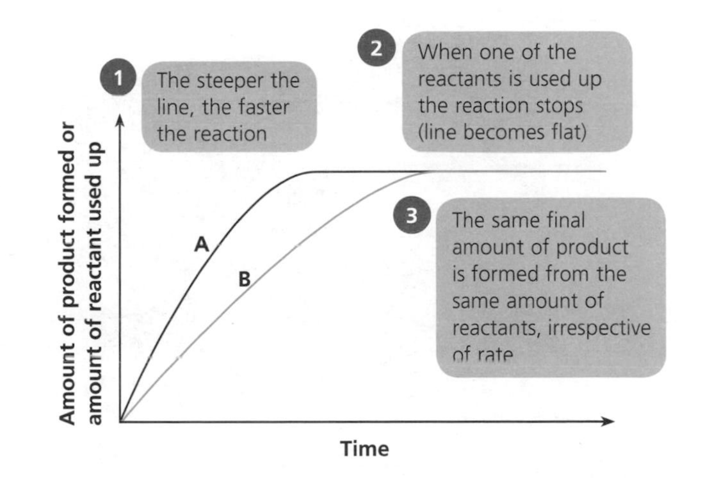

Reaction A is faster than reaction B. This could be because:

- the surface area of the solid reactants in A is greater than in B
- the temperature of reaction A is greater than reaction B
- the concentration of the solution in A is greater than in B
- a catalyst is used in reaction A but not in reaction B.

You need to be able to interpret graphs showing the amount of product formed (or reactant used up) with time, in terms of the rate of the reaction.

The rate of a reaction is the amount of reactant used up or product made, in a given time. One of the factors that affects the rate of reaction is the concentration of dissolved reactants.

An investigation was carried out to find out how different concentrations of hydrochloric acid affect the rate of reaction between marble chips and hydrochloric acid. The reaction is:

calcium carbonate + hydrochloric acid → calcium chloride + carbon dioxide + water

$$CaCO_3(s) + 2HCl(aq) \rightarrow CaCl_2(aq) + CO_2(g) + H_2O(l)$$

The marble chips (calcium carbonate) were in excess in all the reactions.

The table below shows how the original acid was diluted.

Volume of acid (cm^3)	Volume of water (cm^3)	Graph number
0	50	No reaction
10	40	1
20	30	2
30	20	3
40	10	4
50	0	5

Each reaction was started and readings of the volume of carbon dioxide gas collected in a gas syringe were taken continuously over 10 minutes. The volume was recorded every 30 seconds. This method of collecting results or data is called a continuous method. This method measures the rate of the product made.

A graph was plotted of the volume of carbon dioxide produced against time for each concentration of the hydrochloric acid.

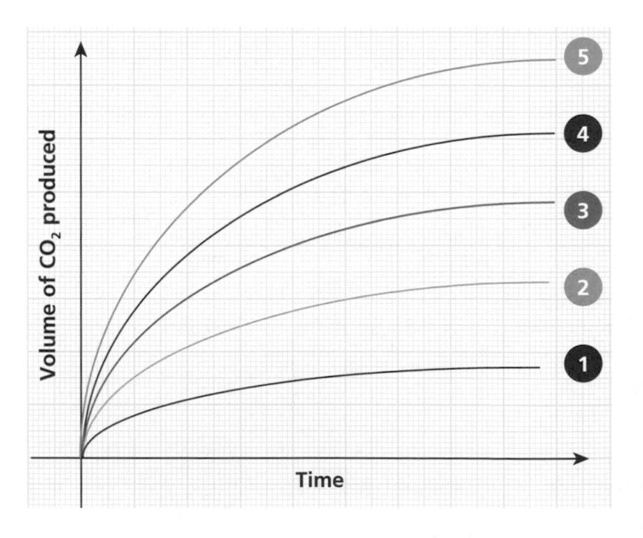

The rate of reaction can be worked out from the graphs. We can see that the most concentrated solution (graph 5) has the steepest curve on the graph, which means it produced CO_2 the quickest. Graph 1 shows the least steep curve, which tells us that this reaction was the slowest at producing CO_2.

You need to be able to explain and evaluate the development, advantages and disadvantages of using catalysts in industrial processes.

Catalysts are substances that speed up the rate of a reaction while remaining chemically unchanged and without being used up, which means they can be used repeatedly.

As this makes the reaction faster, and the product is produced more quickly, it can save an enormous amount of money in labour and energy costs. Greater quantities of the product can then be made quickly to meet further demand.

Catalysts provide an alternative route for a reaction with lower activation energy. This means that less energy is required to start the reaction and it can be carried out at lower temperatures. This reduces the energy requirements of the process, which is good for sustainable development and reduces cost.

Catalysts have been used in industrial processes for many years. Different reactions require different catalysts so a number of catalysts have been developed.

Many transition metals are used as catalysts – e.g. iron is used in the production of ammonia, platinum in the production of nitric acid. Silica/aluminium oxide is used in the production of alkenes (which make plastics).

As technology has become more advanced, scientists are now able to work with smaller and smaller structures, such as nanomaterials. Their very, very small size means they have a large surface area in relation to their size, which makes them ideal to be used as industrial catalysts.

However, it is important to remember that different reactions require different catalysts and that some are very expensive to buy. They also need to be removed from the product and cleaned regularly otherwise they become 'poisoned' and will not work properly.

The reactants need to be purified to minimise this.

Nanomaterials used as industrial catalysts

The table below lists the advantages and disadvantages of using catalysts.

Advantages of Catalysts
• They increase the rate of reaction, which speeds up industrial processes. • They reduce the costs of industrial processes. • They save energy. • They help sustainable development. • They can be used over and over again.
Disadvantages of Catalysts
• Some are expensive. • They need to be separated from the products. • They need to be cleaned regularly to prevent them becoming poisoned.

C2.5 Exothermic and endothermic reactions

Chemical reactions either take in or release energy. To understand this, you need to know:

- what the terms exothermic and endothermic mean
- if a reversible reaction is exothermic in the forward direction, it will be endothermic in the backward direction and vice versa.

When chemical reactions occur, energy is transferred to or from the surroundings, so many chemical reactions are accompanied by a temperature change.

Exothermic Reactions

These reactions are accompanied by a temperature rise. They are known as **exothermic** reactions because they transfer heat energy to the surroundings, i.e. they give out heat. Combustion is a common example of an exothermic reaction:

methane (natural gas) + oxygen → carbon dioxide + water + heat energy
$CH_4(g) + 2O_2(g) \rightarrow CO_2(g) + 2H_2O(l)$

It is not only reactions between fuels and oxygen that are exothermic. Neutralising alkalis with acids gives out heat too, as do many oxidation reactions.

Exothermic reactions are used in everyday life, for example in self-heating cans (for coffee) and in hand warmers.

Endothermic Reactions

These reactions are accompanied by a fall in temperature. They are known as **endothermic** reactions because heat energy is transferred from the surroundings, i.e. they take in heat. Dissolving ammonium nitrate crystals in water is an endothermic reaction:

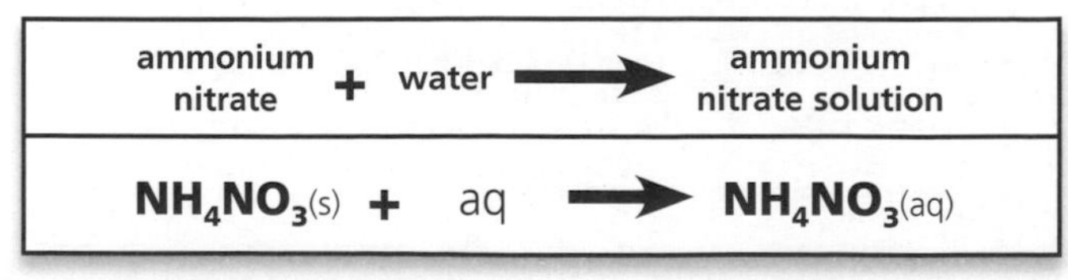

ammonium nitrate + water → ammonium nitrate solution
$NH_4NO_3(s) + aq \rightarrow NH_4NO_3(aq)$

Thermal decomposition is an example of an endothermic reaction. Endothermic reactions are used in everyday life, for example in sports injury packs used for cooling sprains.

Reversible Reactions

If a reaction is reversible and it is exothermic in one direction then it follows that it is endothermic in the opposite direction, with the same amount of energy being transferred in each case. An example of this is when hydrated copper sulfate is heated gently:

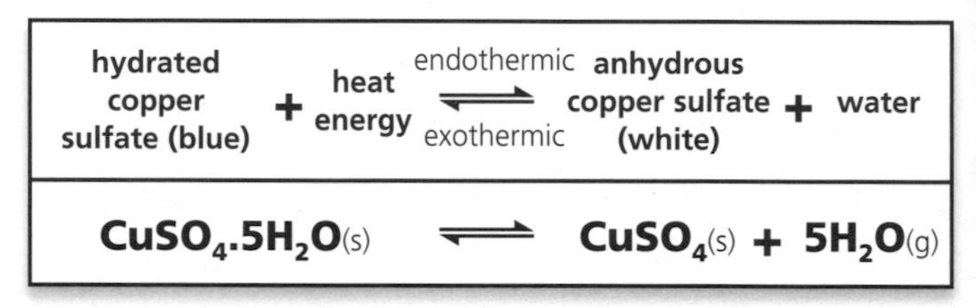

hydrated copper sulfate (blue) + heat energy ⇌ (endothermic / exothermic) anhydrous copper sulfate (white) + water
$CuSO_4.5H_2O(s) \rightleftharpoons CuSO_4(s) + 5H_2O(g)$

Blue crystals of hydrated copper sulfate become white anhydrous copper sulfate on heating, as water is removed.

Hydrated Copper Sulfate **Anhydrous Copper Sulfate**

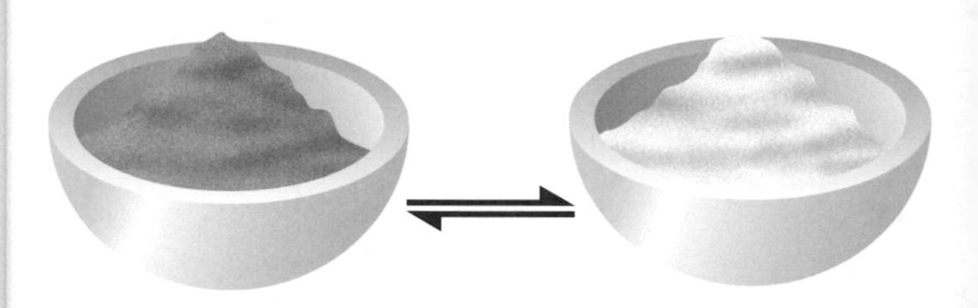

If water is added to white anhydrous copper sulfate, blue hydrated copper sulfate is formed and heat is given out.

This reaction can be used as a test for water, where the colour change from white to blue is an indication of the presence of water.

C2.6 Acids, bases and salts

Soluble and insoluble salts can be made from various chemical reactions. Acids and alkalis react in a neutralisation reaction. To understand this, you need to know:

- how to make soluble and insoluble salts
- what acids and bases are and what the pH scale measures
- the state symbols.

State Symbols

The states of the compounds and elements involved in reactions are shown using the state symbols: (s) solid, (l) liquid, (aq) aqueous solution and (g) gas. An aqueous solution is produced when a substance is dissolved in water.

Soluble Salts

Soluble salts can be made from acids by reacting them with metals, insoluble bases and alkalis (soluble bases).

Some metals react with dilute acid to form a metal salt and hydrogen. **Salt** is a word used to describe many ionic compounds, e.g. the metal compound formed when an acid reacts with a metal.

However, some metals react with acids more vigorously than others:

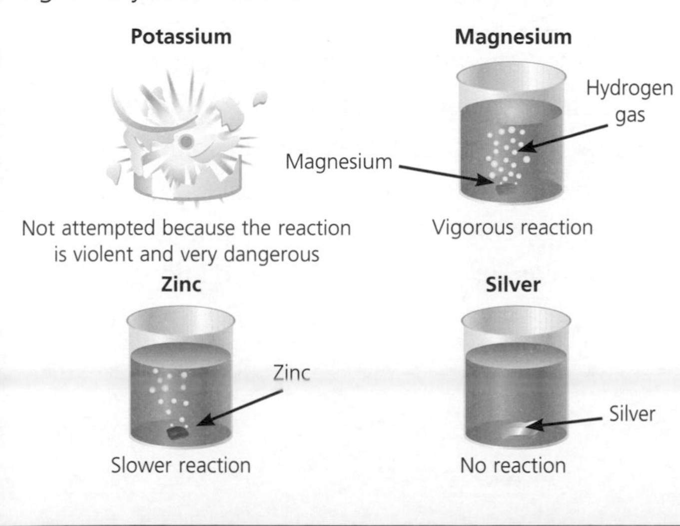

Soluble Salts from Insoluble Bases

Bases are the oxides and hydroxides of metals. Those which are soluble are called **alkalis**.

Unfortunately, the oxides and hydroxides of transition metals are insoluble, which means that preparing their salts is a little less straightforward.

The metal oxide or hydroxide is added to an acid until no more will react. The excess metal oxide or hydroxide is then filtered off to leave a solution of the salt, which can then be evaporated to dryness.

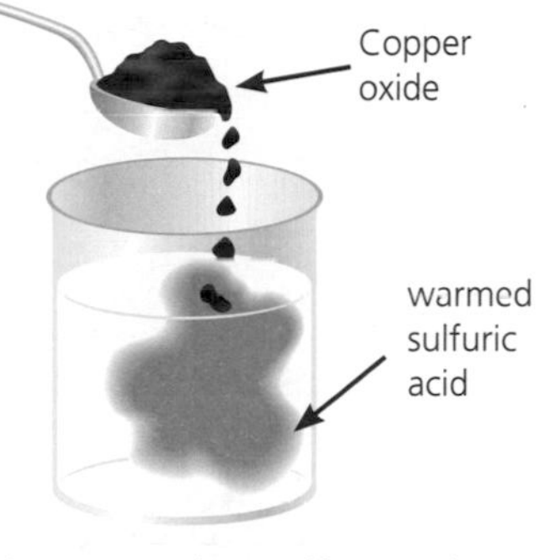

Add copper oxide to sulfuric acid

Filter to remove any unreacted copper oxide

Evaporate to leave behind blue crystals of the 'salt' copper sulfate

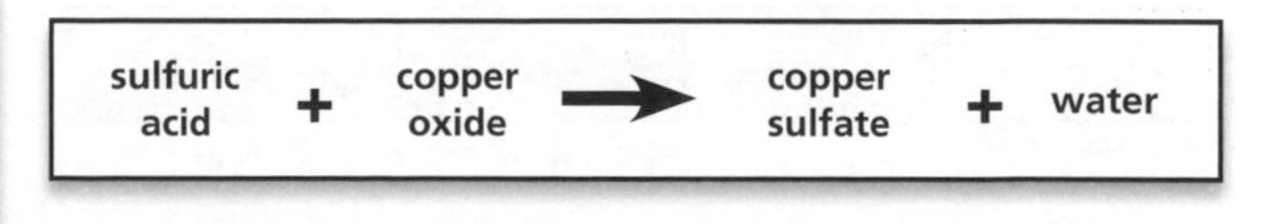

This can be written more generally as:

acid + base ➔ salt + water

Salts from Soluble Bases

Ammonia dissolves in water to produce an alkaline solution. This can be neutralised with acids to produce ammonium salts, which are important as fertilisers.

	Hydrochloric Acid	Sulfuric Acid	Nitric Acid
Ammonium Hydroxide	Ammonium chloride and water	Ammonium sulfate and water	Ammonium nitrate and water

Compounds of alkali metals can be made by reacting solutions of their hydroxides (which are alkaline) with a particular acid. This is called a neutralisation reaction and can be represented as follows:

Acids react with bases/alkalis to produce a salt and water. The salt produced depends on the acid used and the metal in the base/alkali.

Salt names start with the name of the metal in the base/alkali and end with the acid name (altered slightly):

Acid Name	Salt Name Ending
Hydrochloric acid	Chloride
Nitric acid	Nitrate
Sulfuric acid	Sulfate

For example, hydrochloric acid reacting with magnesium oxide would produce the salt called magnesium chloride. Nitric acid reacting with sodium hydroxide would produce sodium nitrate.

	Hydrochloric acid	Sulfuric acid	Nitric acid
Sodium hydroxide	Sodium chloride and water	Sodium sulfate and water	Sodium nitrate and water
Potassium hydroxide	Potassium chloride and water	Potassium sulfate and water	Potassium nitrate and water

Indicators are used to show when the reaction between the acid and alkali is complete. Indicators change colour depending on whether they are in acidic or alkaline solution.

Solid Salts

Once a salt solution has been formed, a solid salt can be produced by heating the solution to evaporate some of the water, which allows the solid salt to crystallise.

Insoluble Salts

Insoluble salts can be made by mixing appropriate solutions of ions so that a solid substance (precipitate) is formed.

Precipitation can be used to remove unwanted ions from solution, e.g. softening hard water. The calcium (or magnesium) ions are precipitated out as insoluble calcium (or magnesium) carbonate.

Acids and Bases

The pH scale is used to measure the acidity or alkalinity of an aqueous solution. Acids have a pH lower than 7 and alkali solutions have a pH greater than 7. Bases are the chemical opposites of acids. Soluble bases are known as alkalis, e.g. sodium hydroxide is an alkali.

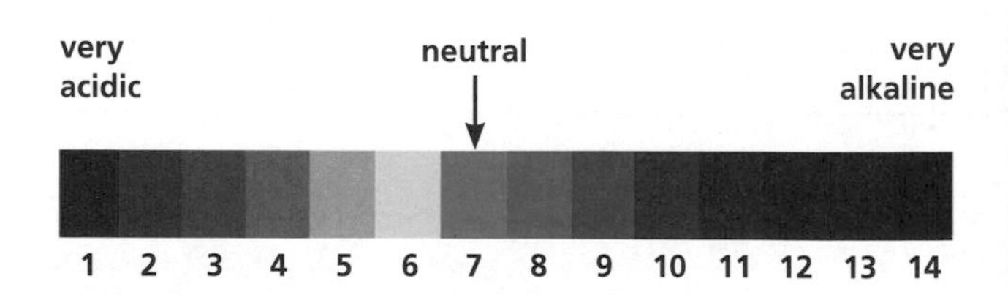

When acids and alkalis dissolve in water they dissociate into their individual ions. All acids dissociate to produce $H^+(aq)$ ions in solution. Alkalis dissociate to produce $OH^-(aq)$ ions in solution.

- Hydrogen ions ($H^+(aq)$) make solutions acidic.
- Hydroxide ions ($OH^-(aq)$) make solutions alkaline.

Neutralisation

Acids and alkalis are **chemical opposites**, so if they are added together in the correct amounts they can 'cancel' each other out. This is because the hydrogen ions react with hydroxide ions to produce water.

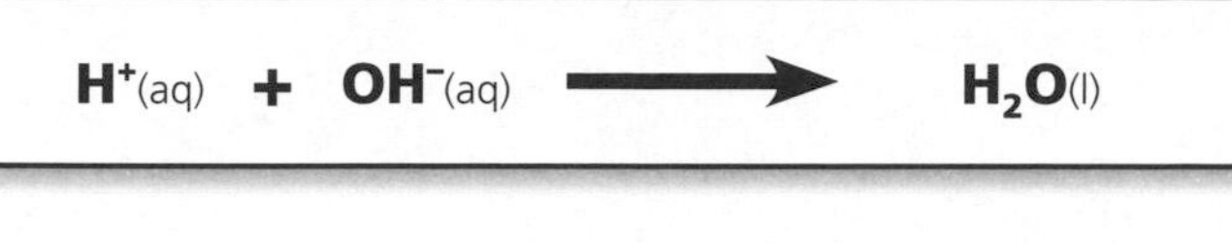

$$H^+(aq) + OH^-(aq) \longrightarrow H_2O(l)$$

This is called neutralisation because the solution that remains has a pH of 7.

acid + alkaline hydroxide solution → neutral salt solution + water

We can see this working if we add the same volumes of HCl (a strong acid) and KOH (a strong alkali) together.

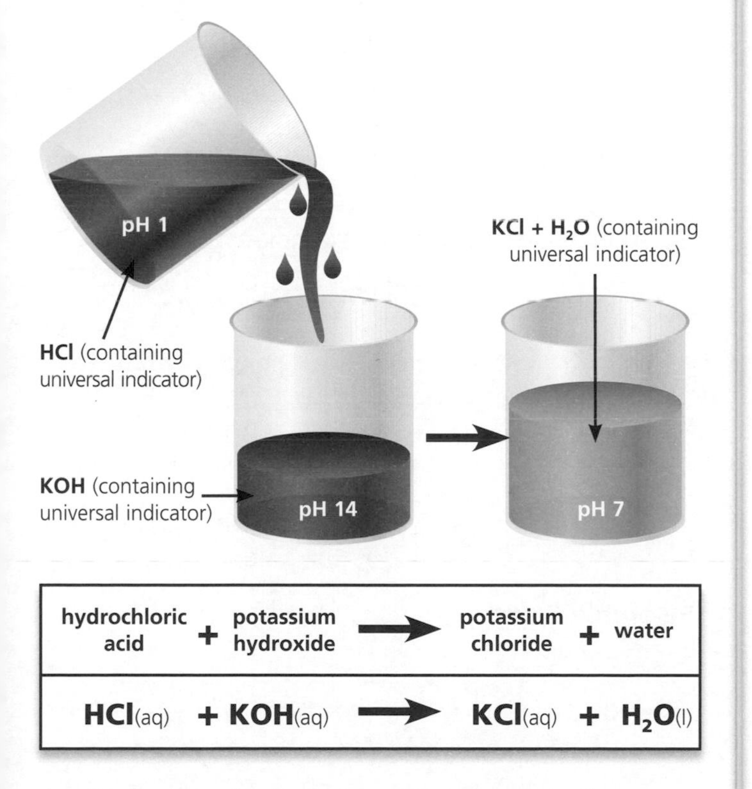

hydrochloric acid	+	potassium hydroxide	→	potassium chloride	+	water
$HCl(aq)$	+	$KOH(aq)$	→	$KCl(aq)$	+	$H_2O(l)$

Again, if we look at what happens to the hydrogen ions $H^+(aq)$ and the hydroxide ions $OH^-(aq)$ in the acid and alkali, we can see that they react to form water:

$$H^+(aq) + OH^-(aq) \longrightarrow H_2O(l)$$

Ammonia is an alkaline gas that dissolves in water to make an alkaline solution. Its main use is in the production of fertilisers to increase the nitrogen content of the soil. Ammonia neutralises nitric acid to produce ammonium nitrate (a fertiliser rich in nitrogen), which is sometimes known as 'nitram' (nitrate of ammonia). The aqueous ammonium nitrate is then evaporated to give the solid.

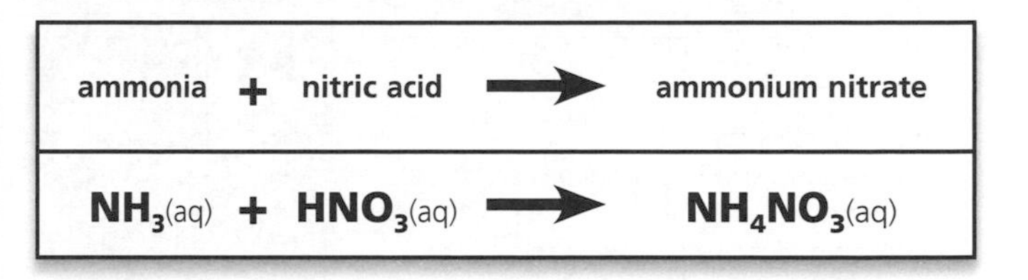

ammonia	+	nitric acid	→	ammonium nitrate
$NH_3(aq)$	+	$HNO_3(aq)$	→	$NH_4NO_3(aq)$

Ammonium nitrate is an example of a widely used nitrogen-based fertiliser. These are important chemicals because they increase the yields of crops. However, nitrates can create problems if they find their way into streams, rivers or groundwater – they can upset the natural balance and contaminate our drinking water.

Nitrate compounds are very soluble in water and are therefore easily washed off fields when it rains. An increase in nitrate levels in reservoirs and lakes can encourage algae growth, which can prevent other plants from getting enough sunlight and therefore causing them to die. This can eventually cause lakes to become lifeless because bacteria break down the dead plants and use up all the oxygen in the water.

If too much nitrate gets into drinking water it can lead to health problems, such as blue baby syndrome. Too much nitrate in drinking water given to small children can stop their blood from carrying enough oxygen. It could also result in long-term respiratory problems.

You need to be able to suggest methods of making a named salt.

Example

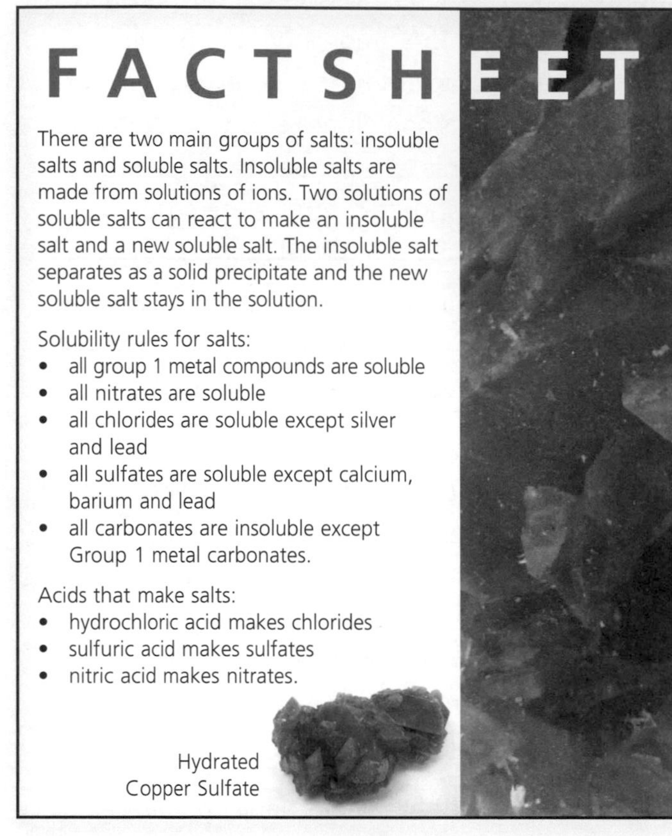

FACTSHEET

There are two main groups of salts: insoluble salts and soluble salts. Insoluble salts are made from solutions of ions. Two solutions of soluble salts can react to make an insoluble salt and a new soluble salt. The insoluble salt separates as a solid precipitate and the new soluble salt stays in the solution.

Solubility rules for salts:

- all group 1 metal compounds are soluble
- all nitrates are soluble
- all chlorides are soluble except silver and lead
- all sulfates are soluble except calcium, barium and lead
- all carbonates are insoluble except Group 1 metal carbonates.

Acids that make salts:

- hydrochloric acid makes chlorides
- sulfuric acid makes sulfates
- nitric acid makes nitrates.

Hydrated Copper Sulfate

Using the information above, suggest which method (shown right) you would use to make copper sulfate and silver chloride and explain your reasoning.

Copper sulfate is a soluble salt. Copper will not react with acids because it is an unreactive metal. The salt cannot be made from the metal and acid, so we can use the metal oxide and acid. Sulfuric acid makes sulfates. The excess solid is then filtered off and the solution of copper sulfate is heated gently to evaporate some of the water. The concentrated solution is then left to crystallise.

Silver chloride is an insoluble salt. The method of making silver chloride could use the soluble salt silver nitrate and the soluble salt sodium chloride. Add solutions of the two soluble salts together and insoluble silver chloride will precipitate. The insoluble precipitate can be removed from the solution of sodium nitrate by filtration. The precipitate is then washed and left to dry at room temperature.

Soluble Salts

1. **Acid + metal**, for example:

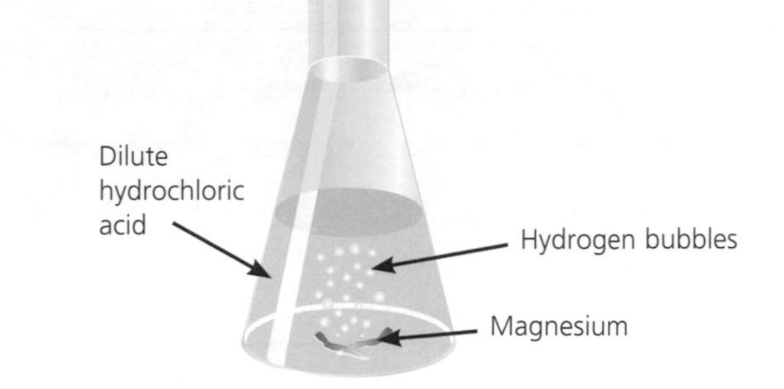

2. **Acid + metal oxide**, for example:

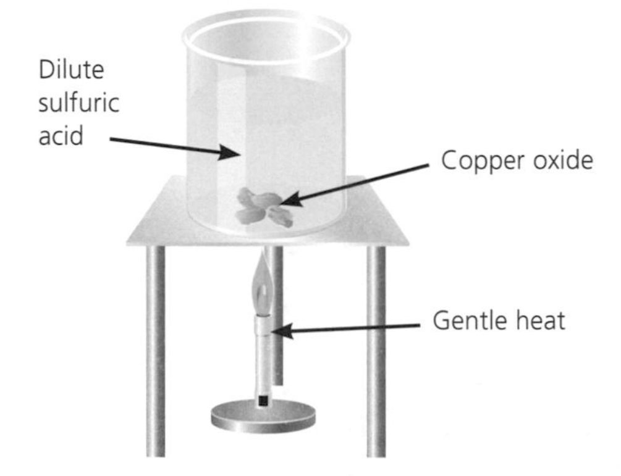

3. **Acid + carbonate**, for example:

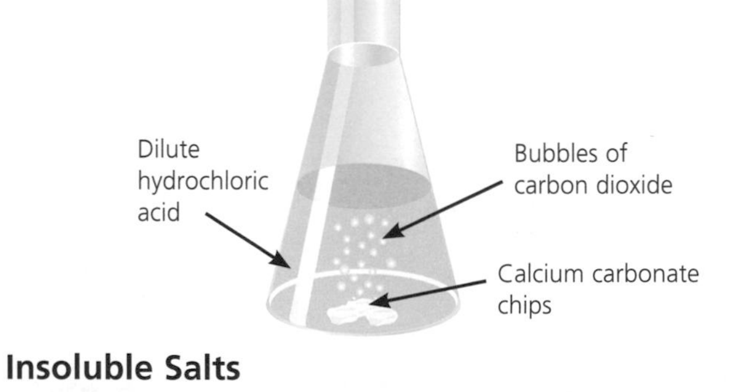

Insoluble Salts

For example:

+ →

Solution of sodium chloride

Solution of silver nitrate

Precipitate of silver chloride

Solution of sodium nitrate

sodium chloride (soluble)	**+**	**silver nitrate** (soluble)	**→**	**silver chloride** (insoluble)	**+**	**sodium nitrate** (soluble)

C2.7 Electrolysis

Electrolysis can be used to produce many useful substances, such as sodium hydroxide, chlorine and hydrogen. To understand this, you need to know:

- the properties of ions in solutions
- what the term electrolysis means
- some uses of electrolysis (extraction of aluminium and electrolysis of sodium chloride solution)
- what reduction and oxidation reactions are.

Electrolysis

When ionic substances are molten (melted) or dissolved in water, the ions are free to move. When an electric current is passed through an ionic substance in solution or a molten ionic substance, it breaks them down into elements. Electrolysis is the breaking down of an ionic compound, making elements, using electricity.

During electrolysis, ions gain or lose electrons at the electrodes forming neutral atoms or molecules. Positively charged ions move to the negative electrode and negatively charged ions move to the positive electrode.

Electrolysis of Copper Chloride ($CuCl_2$)

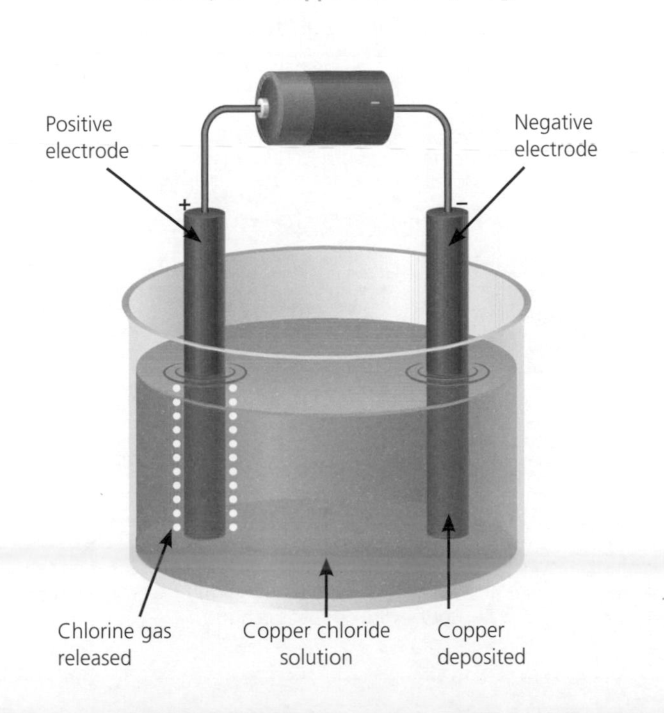

In the diagram the negative chloride ions (Cl^-) move to the positive electrode where they lose electrons to form chlorine atoms (Cl) and then chlorine molecules (Cl_2). The positive copper ions (Cu^{2+}) move to the negative electrode where they gain electrons to form copper atoms (Cu).

Redox Reactions

During electrolysis, positively charged ions gain electrons at the negative electrode. This gain of electrons is known as **reduction**.

At the positive electrode, negatively charged ions lose electrons. This loss of electrons is known as **oxidation**.

A chemical reaction in which both reduction and oxidation occurs is called a **redox** reaction.

It will help you to remember the above if you remember the word **oilrig**.

- **O**xidation **I**s **L**oss of electrons (**OIL**)
- **R**eduction **I**s **G**ain of electrons (**RIG**)

If the electrolyte contains a mixture of ions, the products formed depend on the reactivity of the elements involved.

Extraction of Aluminium

Aluminium is extracted from its ore (bauxite) by electrolysis. The bauxite is purified into aluminium oxide (Al_2O_3) from which the aluminium is extracted.

For electrolysis to take place the aluminium oxide must be molten because the ions are not free to move in the solid state.

Aluminium oxide has a very high melting point and so it is first dissolved in cryolite, which lowers the melting point (from approx 2000°C to 900°C) and allows electrolysis to be carried out at a much lower temperature. This makes it both cheaper and safer.

Extraction of Aluminium (cont)

Electrolysis of Aluminium Oxide

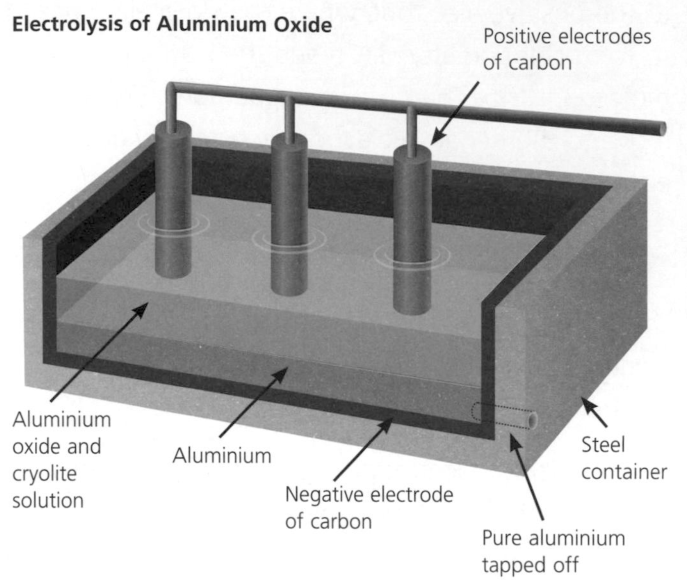

- The electrodes are made of carbon.
- The aluminium ions (Al^{3+}) are attracted to the negative electrode.
- Aluminium (Al) is formed at the negative electrode.
- Aluminium is more dense than the aluminium oxide and cryolite mixture, so it sinks to the bottom of the container where it can be run off and collected.
- The oxide ions (O^{2-}) are attracted to the positive electrode.
- Oxygen is formed at the positive electrode and then reacts with the carbon of the electrode to form carbon dioxide (CO_2).
- As the oxygen reacts with the carbon electrode, the electrode slowly disappears and needs to be replaced.

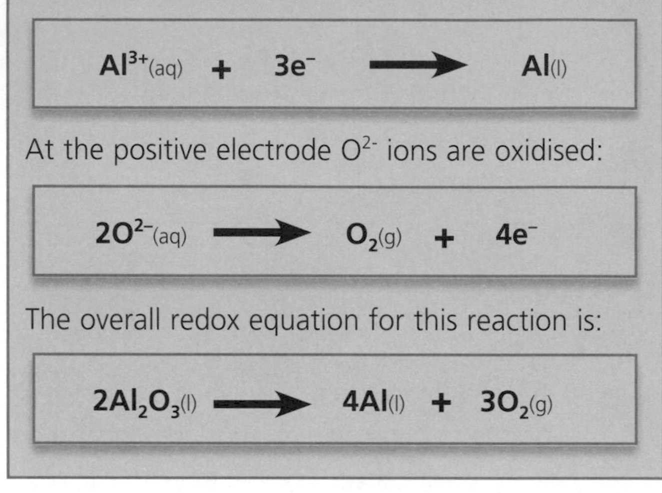

HT The reactions that occur at the electrodes can be represented by half-equations. The following equations show the reactions that occur at the electrodes during the electrolysis of aluminium oxide.

At the negative electrode Al^{3+} ions are reduced:

$$Al^{3+}_{(aq)} + 3e^- \longrightarrow Al_{(l)}$$

At the positive electrode O^{2-} ions are oxidised:

$$2O^{2-}_{(aq)} \longrightarrow O_{2(g)} + 4e^-$$

The overall redox equation for this reaction is:

$$2Al_2O_{3(l)} \longrightarrow 4Al_{(l)} + 3O_{2(g)}$$

Industrial Electrolysis of Sodium Chloride Solution (Brine)

Sodium chloride (common salt) is a compound of an alkali metal and a halogen. It is found in large quantities in the sea and in underground deposits. Electrolysis of sodium chloride solution produces some very important reagents for the chemical industry:

- chlorine gas at the positive electrode
- hydrogen gas at the negative electrode
- sodium hydroxide solution, which is passed out of the cell.

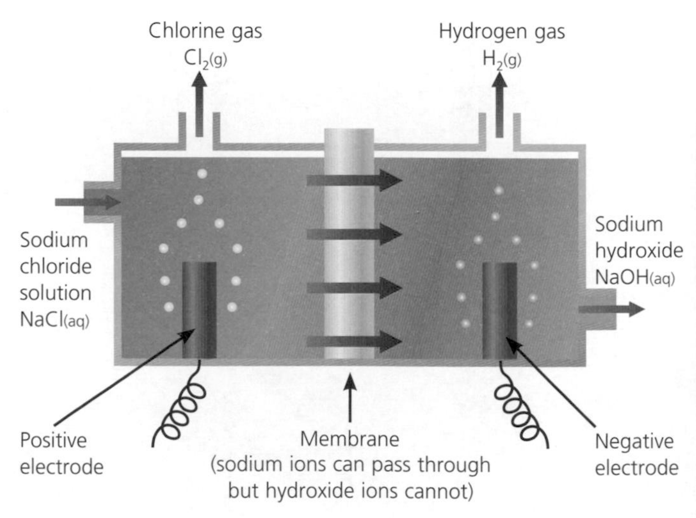

The products of the electrolysis of brine have many uses:

- Chlorine is used to kill bacteria in drinking water and swimming pools and to manufacture hydrochloric acid, disinfectants, bleach and the plastic PVC.
- Hydrogen is used in the manufacture of ammonia and margarine.
- Sodium hydroxide is used in the manufacture of soap, paper and ceramics.

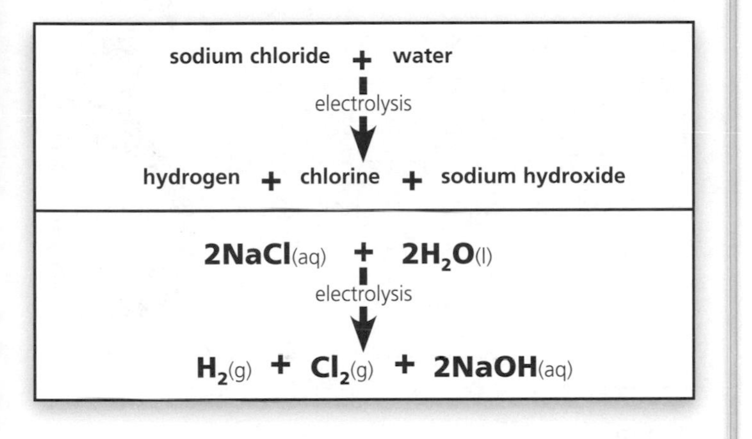

sodium chloride + water $\xrightarrow{\text{electrolysis}}$ hydrogen + chlorine + sodium hydroxide

$$2NaCl_{(aq)} + 2H_2O_{(l)} \xrightarrow{\text{electrolysis}} H_{2(g)} + Cl_{2(g)} + 2NaOH_{(aq)}$$

A simple laboratory test for chlorine is that it bleaches damp litmus paper, i.e. the chlorine removes the colour.

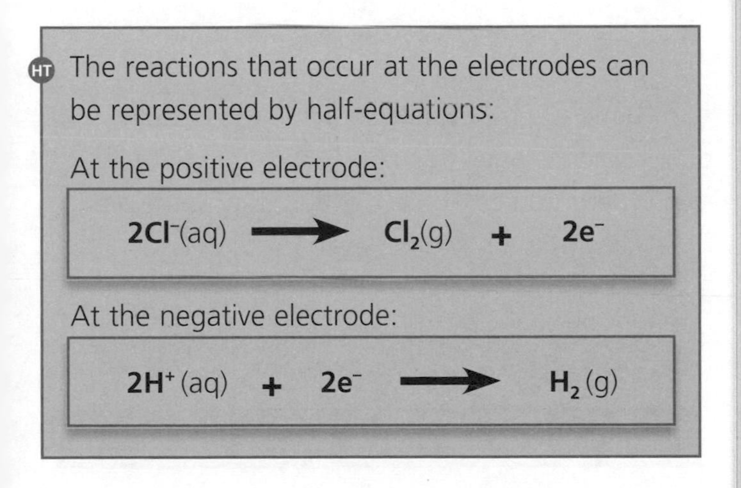

HT The reactions that occur at the electrodes can be represented by half-equations:

At the positive electrode:

$$2Cl^-(aq) \longrightarrow Cl_2(g) + 2e^-$$

At the negative electrode:

$$2H^+(aq) + 2e^- \longrightarrow H_2(g)$$

Other Uses of Electrolysis

Electrolysis is also used to electroplate objects for a variety of reasons, including protecting them from corrosion and improving their appearance. Electroplating involves coating metal objects with layers of other metals. Examples of electroplating include copper plating and silver plating.

Electrolysis can also be used to purify metals, e.g. the purification of copper.

Copper can easily be extracted by reduction of its ore but when it is needed in a pure form it is purified by electrolysis. For electrolysis to take place:

- the positive electrode needs to be made of impure copper
- the negative electrode needs to be made of pure copper
- the solution must contain copper ions.

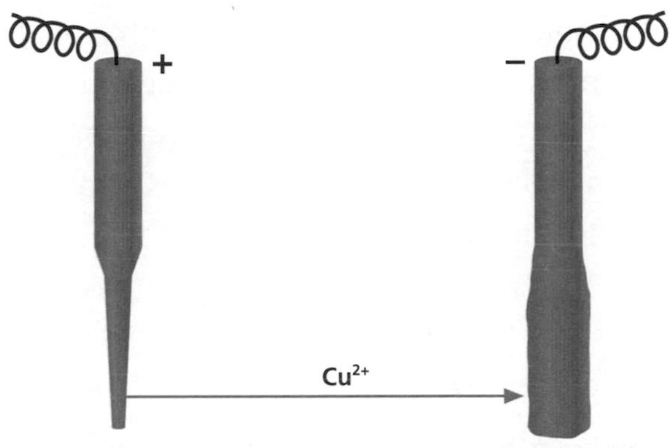

Consequently, the negative electrode gets bigger and bigger, and the positive electrode gradually dissolves away to nothing. The impurities in the positive electrode simply fall to the bottom as the process takes place.

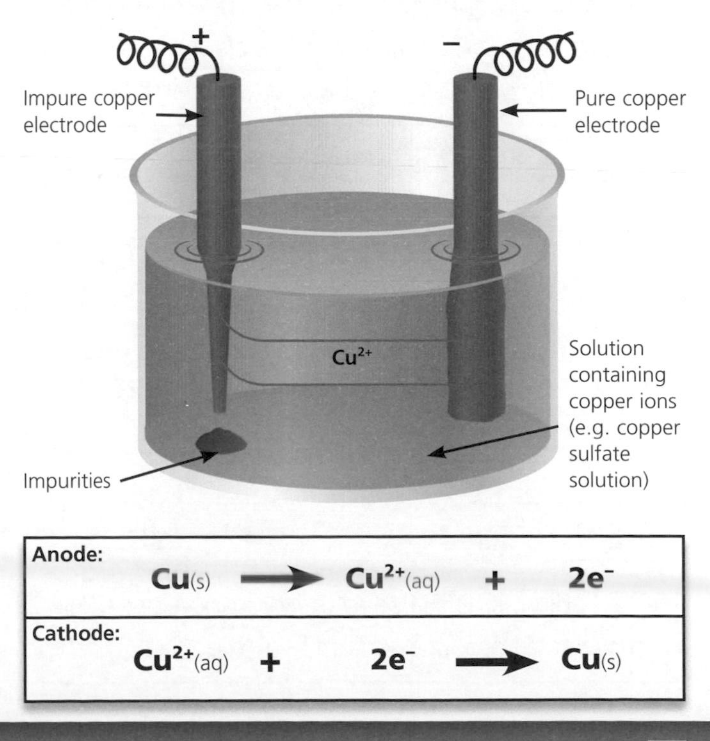

Anode:	$Cu_{(s)} \longrightarrow Cu^{2+}_{(aq)} + 2e^-$
Cathode:	$Cu^{2+}_{(aq)} + 2e^- \longrightarrow Cu_{(s)}$

You need to be able to predict the results of electrolysing solutions of ions.

Ionic substances conduct electricity and can be broken down when they are molten or in solution. The process of electrolysis uses electrical energy to break down these substances into their elements and is used to manufacture important chemical substances that we use in our everyday lives.

The following rules can be used to predict the results of electrolysing solutions:

- If the metal in the solution reacts with acids to make hydrogen, then hydrogen is produced at the negative electrode instead of the metal.
- If the metal does not react with acids to produce hydrogen, then the metal is produced.
- If you have solutions of chlorides, bromides or iodides then chlorine, bromine or iodine is produced at the positive electrode. With other common negative ions, such as sulfate, oxygen is produced.

HT **You need to be able to complete and balance supplied half-equations for the reactions occurring at the electrodes during electrolysis.**

We can write and balance half-equations to show what happens at each electrode during the electrolysis. For example, for the electrolysis of brine (salt water). Write down a word equation:

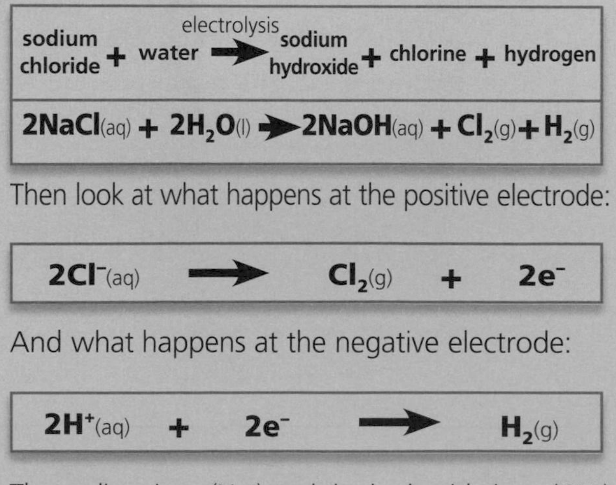

sodium chloride + water $\xrightarrow{\text{electrolysis}}$ sodium hydroxide + chlorine + hydrogen

$$2NaCl_{(aq)} + 2H_2O_{(l)} \rightarrow 2NaOH_{(aq)} + Cl_{2(g)} + H_{2(g)}$$

Then look at what happens at the positive electrode:

$$2Cl^-_{(aq)} \rightarrow Cl_{2(g)} + 2e^-$$

And what happens at the negative electrode:

$$2H^+_{(aq)} + 2e^- \rightarrow H_{2(g)}$$

The sodium ions (Na^+) and the hydroxide ions (OH^-) combine to produce sodium hydroxide (NaOH).

You need to be able to explain and evaluate processes that use the principles described in this unit.

The process of electrolysis uses electrical energy to break down ionic substances, which are molten or in solution, into elements. This process is important industrially because it is used to manufacture important chemical substances we need for our everyday lives.

Example

The electrolysis of sea water or brine gives us large quantities of hydrogen gas, chlorine and sodium hydroxide solution. These products are also used as the starting point for other useful products such as disinfectants, bleach and soap.

The process is very expensive because of all the electrical energy that is required to bring about the change.

Advantages
• Three very important materials (hydrogen, chlorine and sodium hydroxide) are produced from one raw material (brine). • Brine is a renewable source that is readily available and cheap. • The products of the electrolysis of brine are used in many industries to make a large variety of products. • As all the products are useful there is minimal waste.
Disadvantages
• The process of electrolysis is very expensive because it depends on electrical energy to work. • The production of electrical energy in traditional power stations adds to the pollution in the atmosphere. • Hydrogen and chlorine can be produced by other methods much more cheaply.

1. Hydrogen and oxygen react together to form water. The electron arrangement of hydrogen and oxygen are shown here:

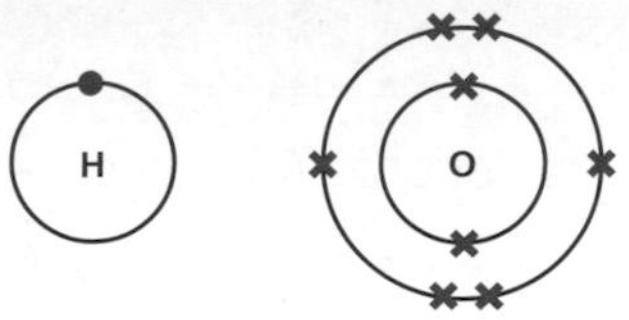

H

O

   a) Draw a diagram to show the bonding in water. You should only use outer shell electrons in your diagram. **(2 marks)**

   b) What type of bonding is formed when hydrogen reacts with oxygen to make water? **(1 mark)**

2. Sodium chloride conducts electricity under certain circumstances. State and explain when sodium chloride will conduct electricity. **(2 marks)**

3. A student reacted magnesium ribbon with hydrochloric acid. This reaction can occur at different rates when the conditions are altered.

   a) Write a word equation for the reaction. **(2 marks)**

   b) The experiment was repeated with magnesium powder rather than magnesium ribbon. The rate of reaction was much faster with magnesium powder than magnesium ribbon. Explain why. **(2 marks)**

   c) The student wanted to increase the rate of reaction even more. Suggest two things the student could alter in order to speed up the reaction further. **(2 marks)**

4. Explain what is meant by the terms 'oxidation' and 'reduction' using 'electrons' in your answer. **(2 marks)**

5. Electrolysis can be used to extract aluminium metal from its ore (bauxite).

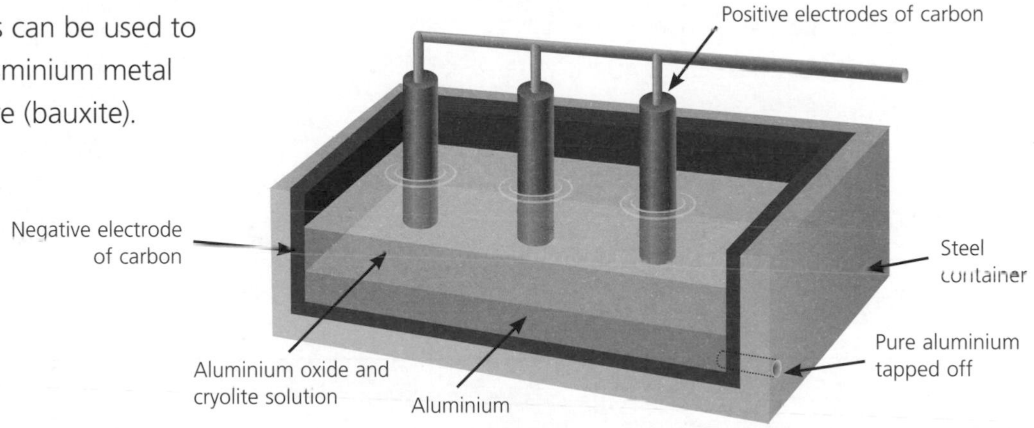

Positive electrodes of carbon

Negative electrode of carbon

Steel container

Pure aluminium tapped off

Aluminium oxide and cryolite solution

Aluminium

   a) Before the aluminium can be extracted, cryolite is added. Explain why cryolite is added to the aluminium ore. **(1 mark)**

   b) What are the electrodes made of in the electrolysis of aluminium? **(1 mark)**

   c) Why does the positive electrode need to be replaced regularly? **(1 mark)**

HT

6. In the electrolysis of bauxite (Al_2O_3) aluminium and oxygen are produced. Write half-equations to show what happens at:

   a) the negative electrode **(2 marks)**

   b) the positive electrode. **(2 marks)**

C3 The Periodic Table

C3.1 The Periodic Table

The Periodic Table groups elements and helps us to understand and predict the properties and reactions of the elements in a group once we know how one element in the group behaves. To understand this, you need to know:

- about early attempts to arrange the elements in the Periodic Table
- how elements are arranged in the modern Periodic Table
- about trends in Groups 1, 7 and the transition elements.

The Periodic Table

In the modern Periodic Table, the elements are arranged in order of **atomic number**. They are then arranged in rows, or **periods**, so that elements with similar properties are in the same columns, or **groups**.

Arranging them in strict order of relative atomic mass would result in some oddities, such as argon ending up in Group 1 and potassium in Group 0, instead of the other way round.

The **Periodic Table** is an arrangement of the elements in terms of their electron structure.

Elements in the same group have the same number of electrons in their outermost shell (except helium). This number also coincides with the group number. Elements in the same group have similar properties.

From left to right across each **period**, a particular energy level is gradually filled with electrons. In the next period, the next energy level is filled, etc. The period number is the same as the number of shells of electrons.

Fewer than a quarter of the elements are non-metals. They are found in the groups to the right-hand side of the Periodic Table.

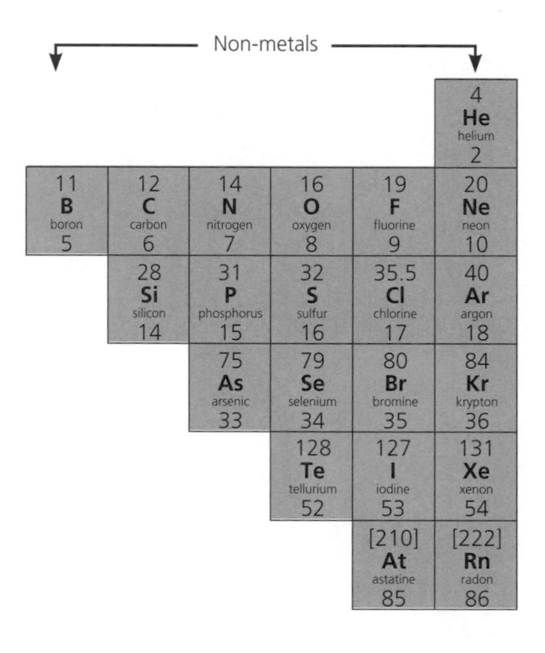

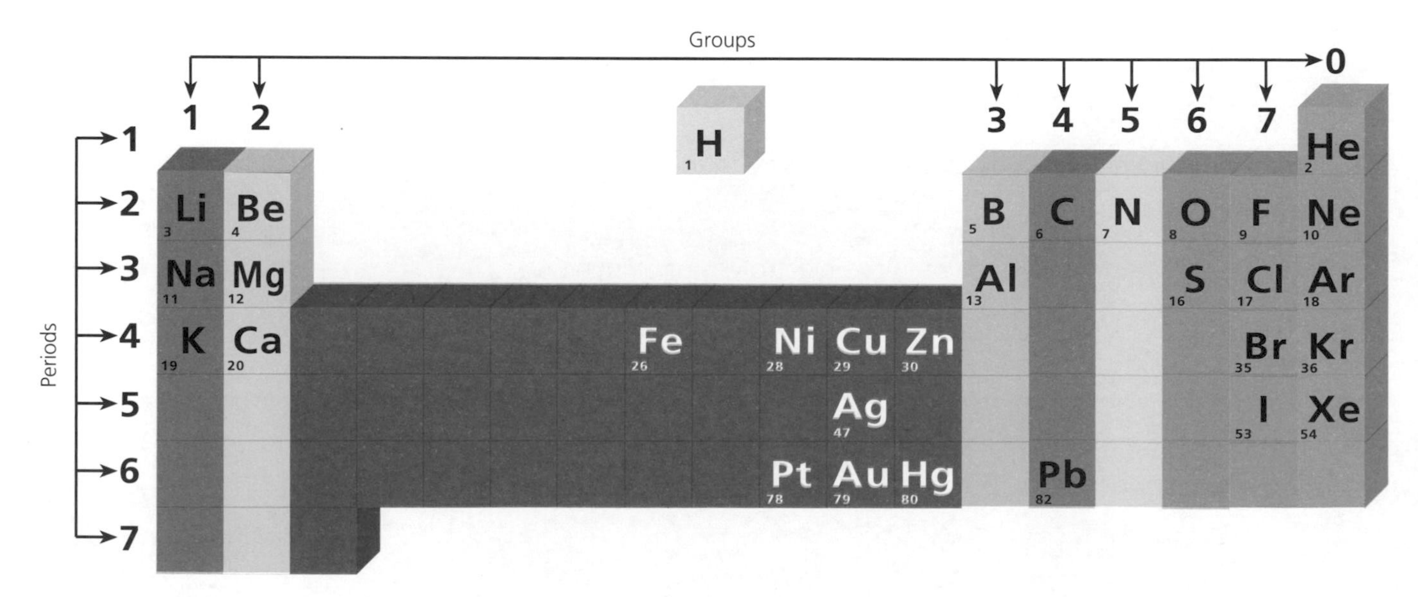

Early Attempts to Classify the Elements

John Newlands (1864)
Newlands knew of the existence of only 63 elements; many were still undiscovered. He arranged the known elements in order of atomic weight and found similar properties amongst every eighth element in the series. This makes sense since the first of the noble gases (Group 0) was not discovered until 1894.

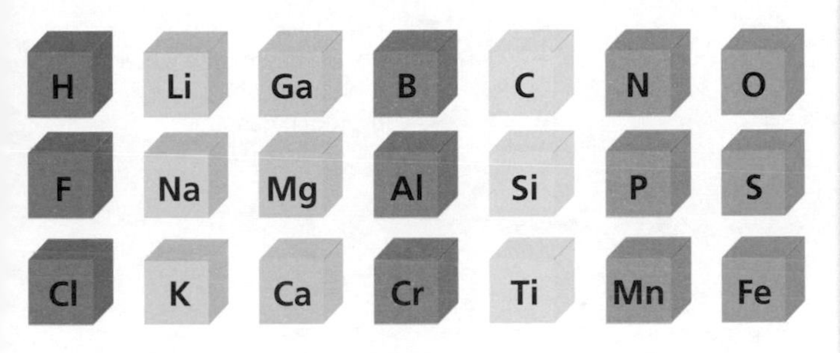

He had noticed periodicity (regular trends along the periods) of the elements he arranged, although the missing elements caused problems and his discovery was not seen as significant.

However, he had not left any spaces in which to insert any new elements yet to be discovered. His arrangement of the elements also meant putting non-metals (oxygen and sulfur) in the same group as a metal (iron). Because of this his table was not accepted by other scientists.

Dimitri Mendeleev (1869)
Mendeleev realised that some elements had yet to be discovered, so he left gaps to accommodate their eventual discovery.

Mendeleev is usually regarded as the founder of the modern Periodic Table. His table of 1869 can be represented as shown in the next column.

He used his periodic table to predict the existence and properties of elements that had yet to be discovered. Scientists were more inclined to accept his table once these elements were discovered and were found to have properties similar to those that Mendeleev predicted.

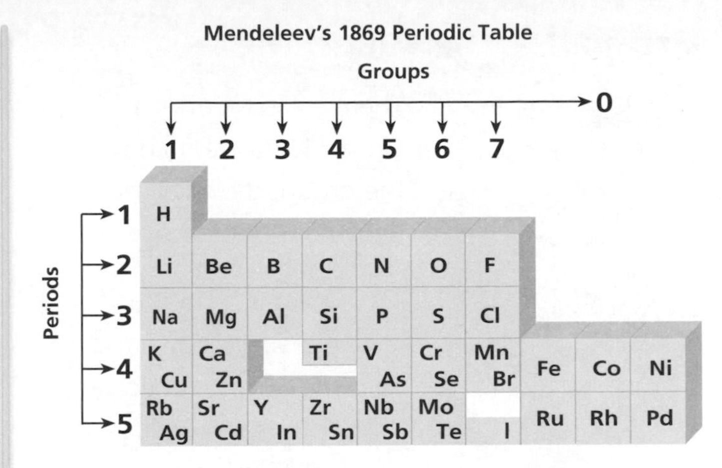

Evidence to support Mendeleev's Periodic Table
Mendeleev's periodic table began to be more widely accepted by scientists when evidence was found to support his arrangement of the elements.

Mendeleev was able to predict the atomic weights of some of the elements he left gaps for and he also predicted their properties.

Shortly after Mendeleev arranged his table, an element called gallium was discovered (1875). Mendeleev had predicted the properties of this undiscovered element in his table and had left a space for it below aluminium. It was found that its properties were very similar to those Mendeleev had predicted, and other predicted elements were also discovered later. These discoveries lent more credit to his version of the periodic table.

Modern Chemistry

Although initially scientists regarded Mendeleev's periodic table as a curiosity, it later became a useful tool.

The discovery of subatomic particles (protons, electrons and neutrons) and electron structure provided a more sound basis for the table because the key to similarities among elements is the number of electrons in the outermost energy level (Group 1 elements have 1 electron in their outermost energy level, Group 2 elements have 2 electrons and so on).

Group 1 – The Alkali Metals

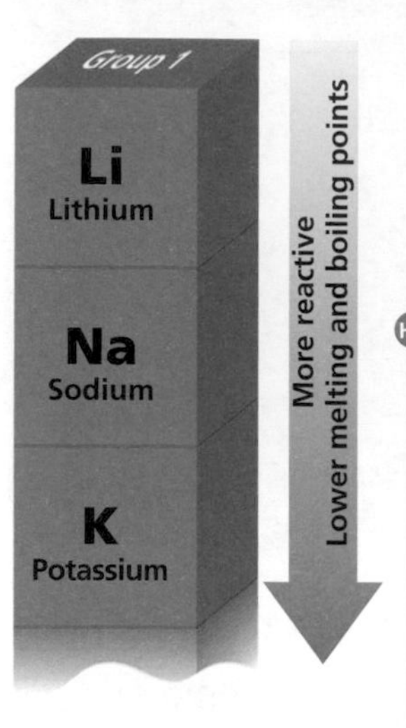

There are six metals in Group 1. As we go down the group, the alkali metals become more reactive.

HT The elements become more reactive down the group because it becomes easier for them to lose their outermost electron. This is due to the atoms increasing in size going down the group.

Alkali metals have low melting points. The melting and boiling points decrease as we go down the group.

Reaction of Alkali Metals with Water

The alkali metals are stored under oil because they react very vigorously with oxygen and with water. Lithium, sodium and potassium float on cold water (because of their low density) and melt because the heat from the reaction is great enough to turn them into liquids. Lithium reacts quite gently, sodium more vigorously and potassium so aggressively that it melts and catches fire. The alkali metals become more reactive as you go down the group. This means potassium is more reactive than sodium.

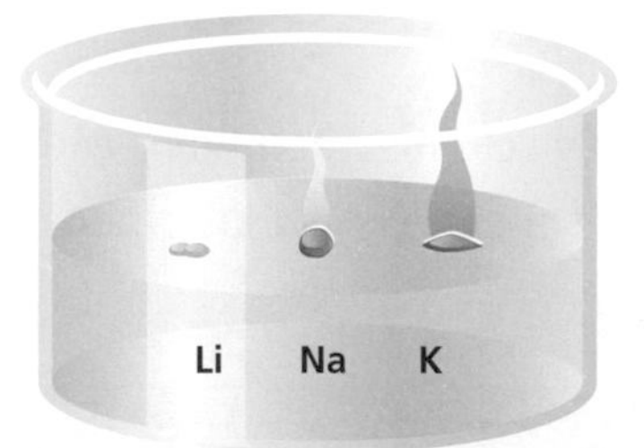

When alkali metals react with water a metal hydroxide and hydrogen gas are formed. The metal hydroxide (e.g. potassium hydroxide) dissolves in water to form an alkaline solution. For example:

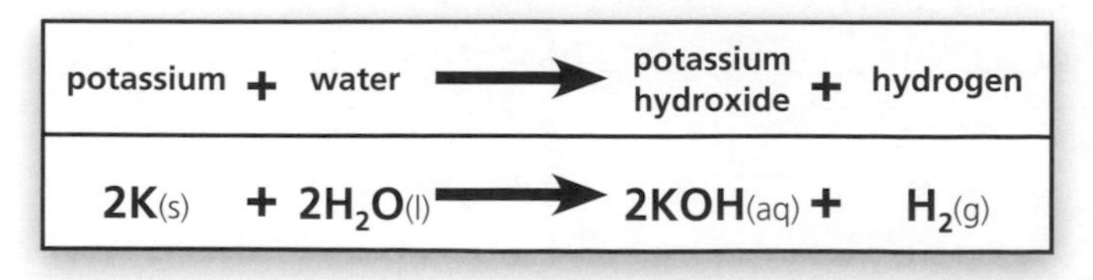

potassium	+	water	→	potassium hydroxide	+	hydrogen
$2K_{(s)}$	+	$2H_2O_{(l)}$	→	$2KOH_{(aq)}$	+	$H_{2(g)}$

Follow these steps to check the pH of the solution formed when an alkali metal is added to water.

1

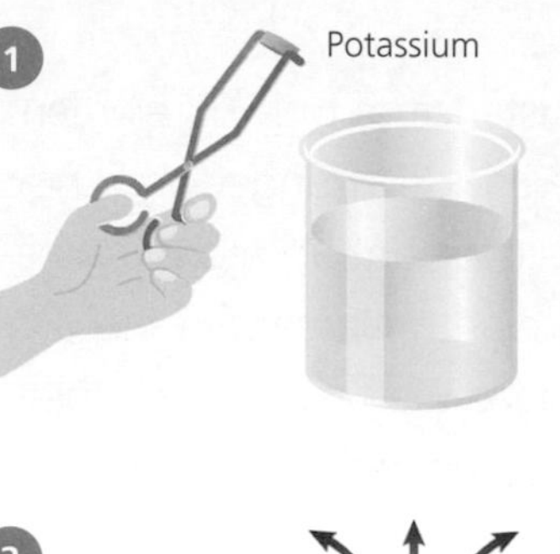

Put some universal indicator in a beaker containing water – the indicator should be green to show neutral pH (pH 7).

2

Put a small piece of potassium into the water. It will react and give off hydrogen gas.

3

When it has finished reacting, the beaker will contain potassium hydroxide solution, $KOH_{(aq)}$. The solution will now be purple, which indicates it is alkaline.

A simple laboratory test for hydrogen gas is to hold a lighted splint to a test tube of the gas. If hydrogen is present, it will burn with a squeaky pop and the flame goes out.

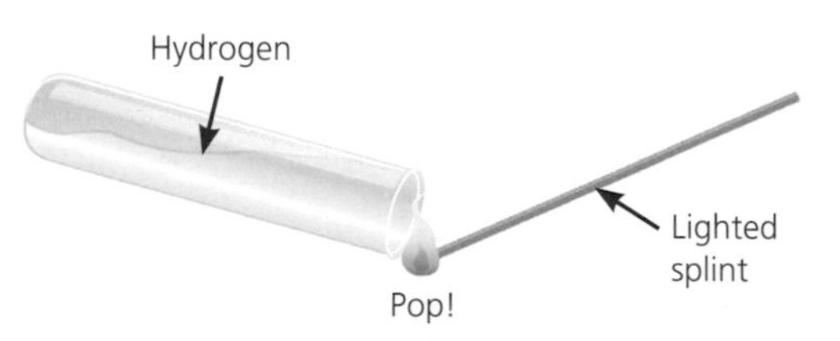

Reaction of Alkali Metals with Non-metals

When alkali metals react with non-metals to form ionic compounds, the metal atoms lose one electron each to form metal ions with a positive charge. The products are white solids that dissolve in water to form colourless solutions. For example:

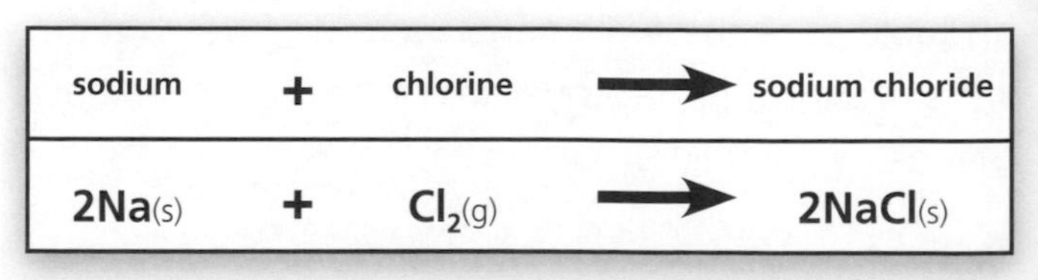

sodium	+	chlorine	→	sodium chloride
$2Na_{(s)}$	+	$Cl_{2(g)}$	→	$2NaCl_{(s)}$

Group 7 – The Halogens

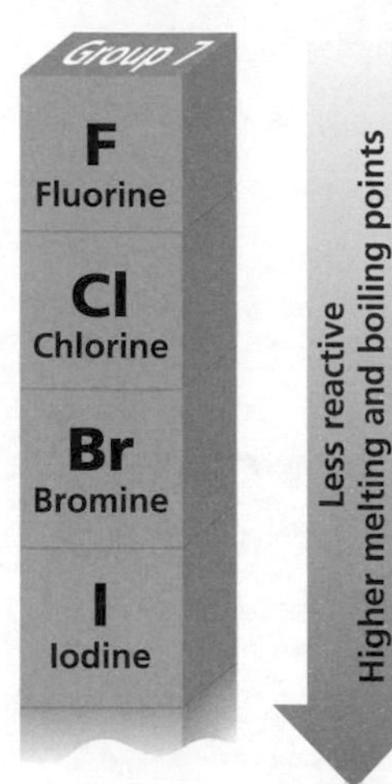

The halogens are non-metal elements. As we go down Group 7 the melting and boiling points increase (fluorine is a gas at room temperature whereas iodine is a solid). The Group 7 elements become less reactive down the group.

HT

The elements become less reactive down the group because it becomes harder for them to gain an electron. This is due to the atoms increasing in size going down the group.

At room temperature, fluorine and chlorine are gases and bromine is a liquid. They all have coloured vapours which, in the case of chlorine and bromine, are extremely pungent.

They exist as molecules made up of pairs of atoms (diatomic).

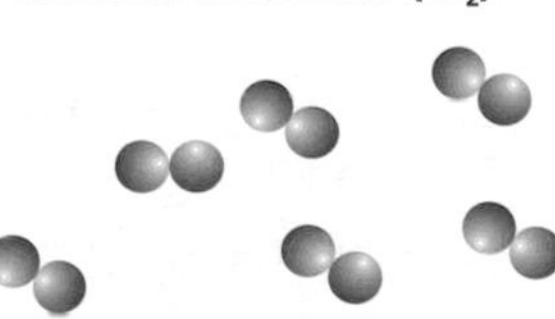

Halogens are brittle and crumbly when solid and are very poor conductors of heat and electricity.

Reaction of Halogens with Metals

Halogens react with metals to produce ionic salts. The halogen atoms gain one electron each to form halide ions with a charge of –1, for example:

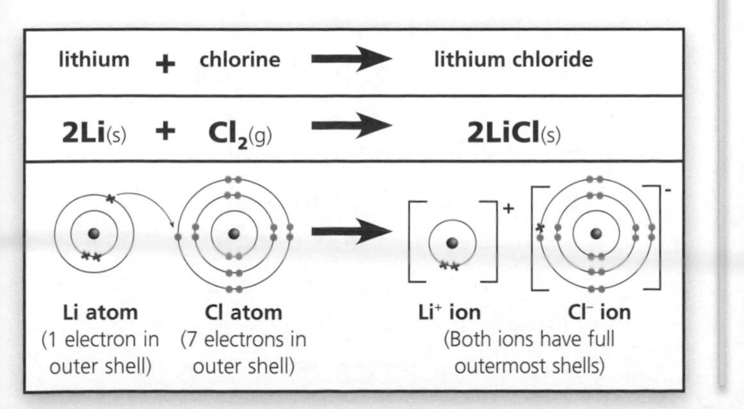

lithium + chlorine → lithium chloride

$2Li_{(s)} + Cl_{2(g)} \rightarrow 2LiCl_{(s)}$

Reaction of Halogens with other Non-metallic Elements

Halogens react with other non-metallic elements to form molecular compounds.

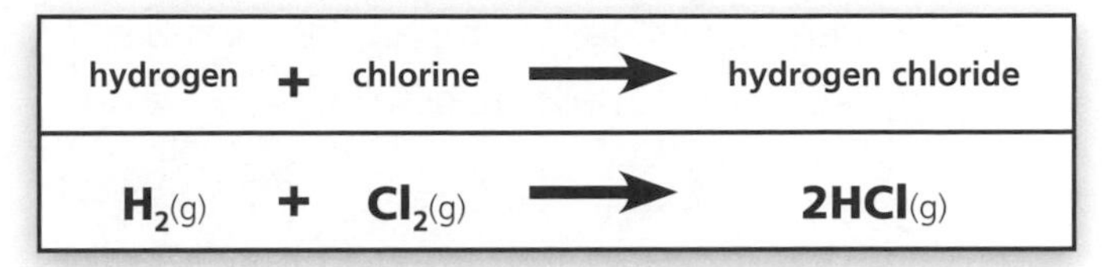

hydrogen + chlorine → hydrogen chloride

$H_{2(g)} + Cl_{2(g)} \rightarrow 2HCl_{(g)}$

Displacement Reactions of Halogens

A more reactive halogen will displace a less reactive halogen from an aqueous solution of its salt – chlorine will displace both bromine and iodine but bromine will displace iodine and not chlorine.

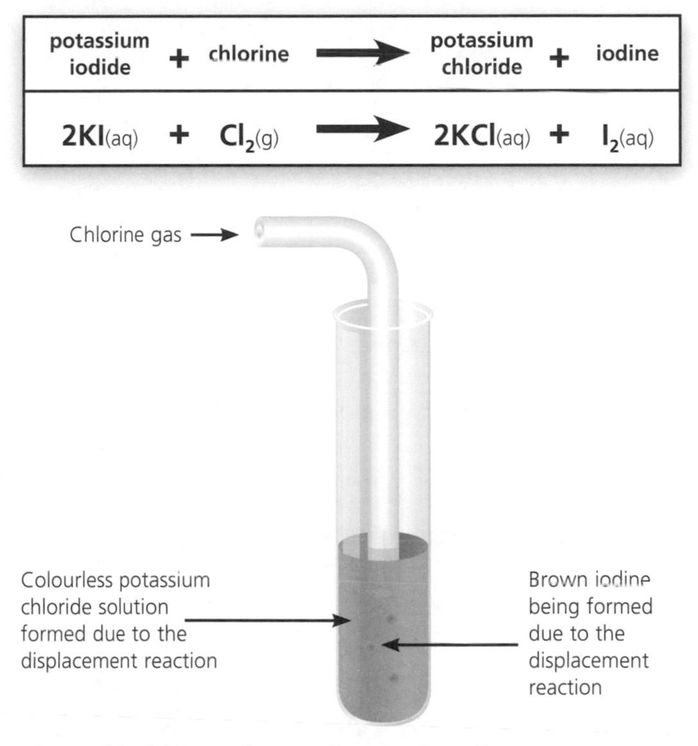

potassium iodide + chlorine → potassium chloride + iodine

$2KI_{(aq)} + Cl_{2(g)} \rightarrow 2KCl_{(aq)} + I_{2(aq)}$

The table below shows the results of reactions between halogens and aqueous solutions of salts.

	Potassium chloride (KCl)	Potassium bromide (KBr)	Potassium iodide (KI)
Chlorine Cl_2	×	Potassium chloride + bromine	Potassium chloride + iodine
Bromine Br_2	No reaction	×	Potassium bromide + iodine
Iodine I_2	No reaction	No reaction	×

Trends in Group 1

Alkali metals all have similar properties because they have the same number of electrons in their outermost shell, i.e. the highest occupied energy level contains one electron.

They become more reactive going down the group, because the outermost electron shell gets further away from the influence of the nucleus and so the outer electron is lost more easily.

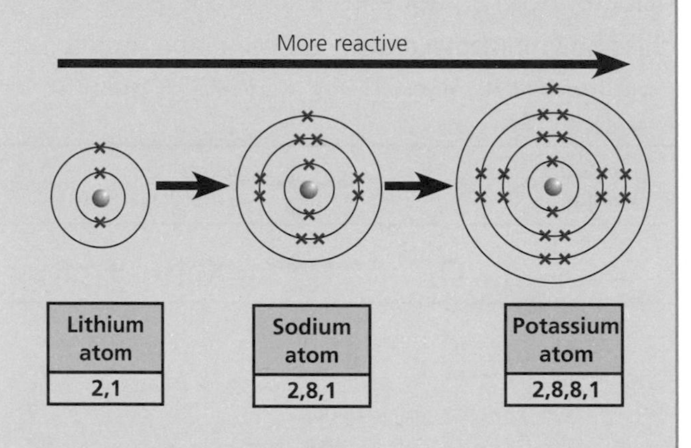

Trends in Group 7

Halogens all have similar properties because they have the same number of electrons in their outermost shell, i.e. the highest occupied energy level contains seven electrons.

They become less reactive going down the group, because the outermost electron shell gets further away from the influence of the nucleus and so an electron is less easily gained.

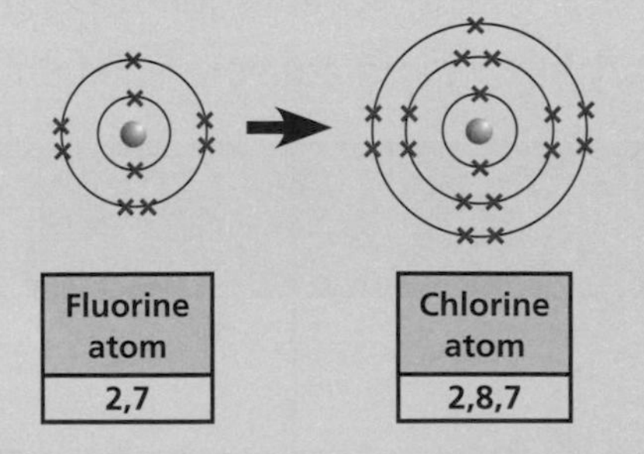

The higher the energy level (i.e. the more shells an atom has):

- the more easily electrons are lost
- the less easily electrons are gained.

The Transition Elements

In the centre of the Periodic Table, between Groups 2 and 3, is a block of metallic elements called the **transition elements**. These include iron, copper, platinum, mercury, chromium and zinc.

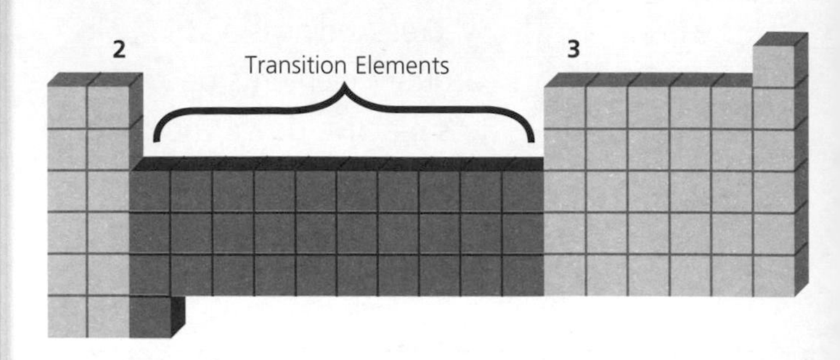

Transition elements are metals and, like all other metals, are good conductors of heat and electricity. They can also be easily bent and hammered into shape. These properties make transition elements very useful as structural materials, and as electrical and thermal conductors.

Although they are all metals, the transition elements have some different properties from the alkali metals of Group 1:

- They have higher melting points (except mercury).
- They are more dense.
- They are stronger and harder (except mercury).
- They are much less reactive, e.g. they do not react as vigorously with water or oxygen. (Copper can be used to make water pipes because it does not react with water).

Although they behave like all the other metals in the Periodic Table, the transition metals also have some properties that are specific to them:

- The elements have ions with different charges, e.g. you can have Fe^{2+} and Fe^{3+} whereas all ions in Group 1 have a +1 charge.
- They form coloured compounds (all Group 1 and 2 compounds are white/colourless).
- They are useful as catalysts, e.g. iron is used as a catalyst in the Haber process.

You need to be able to explain how attempts to classify elements in a systematic way, including those of Newlands and Mendeleev, have led, because of the growth of chemical knowledge, to the modern Periodic Table and explain why scientists regarded the Periodic Table of the elements first as a curiosity, then as a useful tool and finally as an important summary of the structure of atoms.

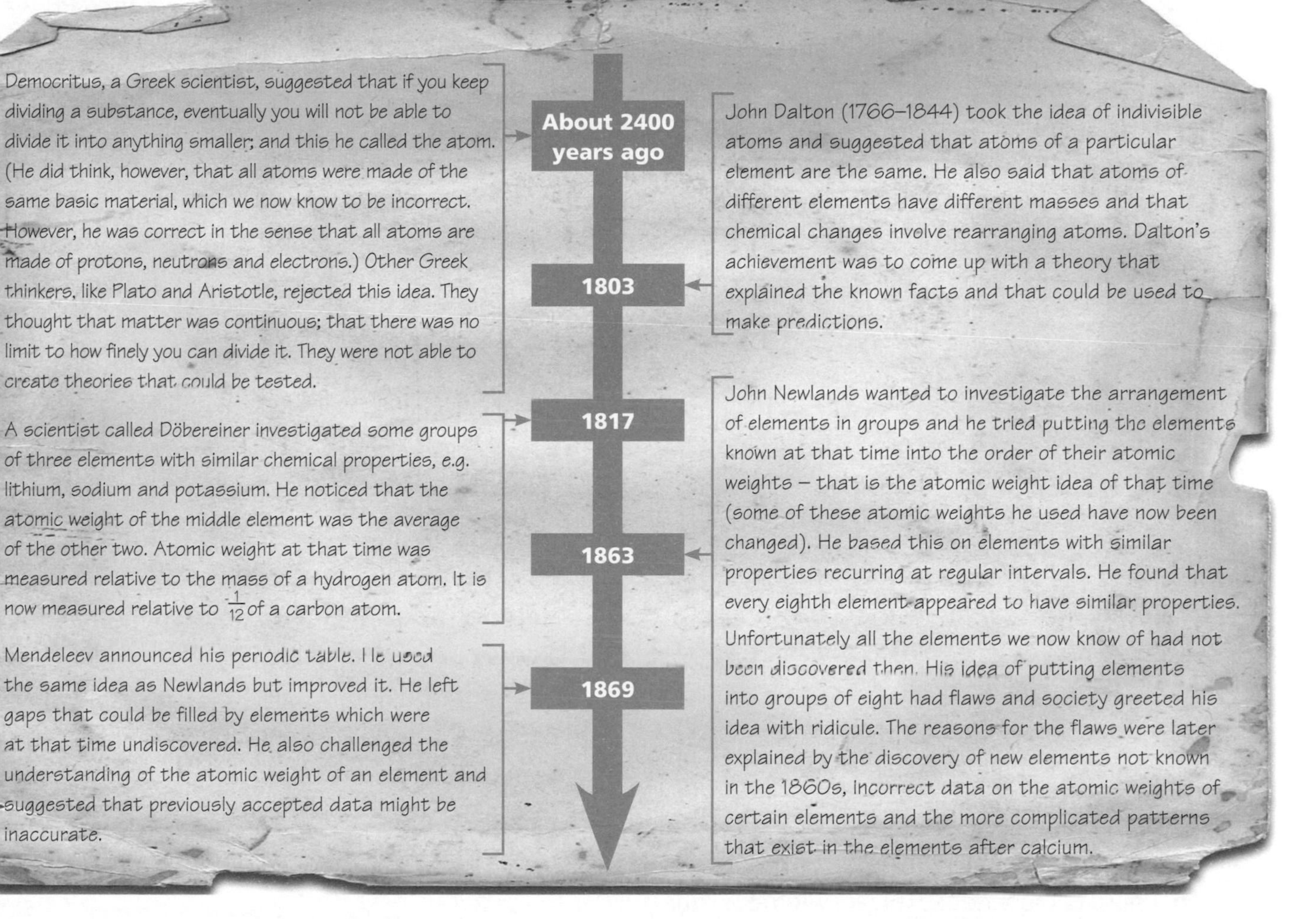

Scientists started to contemplate the idea of atoms 2400 years ago and there have since been many proposed theories.

As knowledge of chemical facts grew in the 18th and 19th centuries, scientists had to find patterns to avoid being overwhelmed by the mass of information and to provide a basis for understanding the facts. Previous knowledge of atomic structure helped scientists to understand elements.

The ideas and evidence of Mendeleev formed the basis of the modern Periodic Table. The elements are arranged in order of their atomic number, not in order of their relative atomic mass.

Research is still taking place to try to find new elements and they are still being discovered, although many of the new elements are radioactive and exist only for a very short time. Their place in the Periodic Table, however, enables predictions to be made about their chemical properties.

Atomic number relates to the number of protons in the nucleus, and it is the number of electrons in the outer shell that governs how an element reacts. Each group in the Periodic Table has elements in it with the same number of electrons in their outer shell. This means elements in the same group of the Periodic Table react in a similar way. They have similar chemical properties. It is possible to predict reactions of elements using evidence from other elements in the same group.

C3.2 Water

The water we drink contains dissolved substances, i.e. it is not pure. Drinking water contains some substances that are beneficial to our health and, because it has been treated, it should not contain anything harmful. Some of the substances in water cause hardness.
To understand this, you need to know:
- the properties of hard and soft water
- how to distinguish between the two types of hard water
- how and why water is purified.

Hard and Soft Water

Water is a solvent and many compounds can dissolve in it. The amount of certain compounds present determines whether the tap water is described as **hard** or **soft**.

Most hard water contains calcium or magnesium compounds which dissolve in natural water that flows over ground or rocks containing compounds of these elements. These dissolved substances react with soap to form scum, which makes it harder to form a lather. Soapless detergents do not form scum.

Soft water does not contain many dissolved calcium or magnesium compounds so it readily forms a lather with soap.

The advantages and disadvantages of hard water are:

Advantages	Disadvantages
• The dissolved compounds in hard water are good for your health, e.g. calcium compounds help in the development of strong bones and teeth. They also help to reduce the development of heart disease.	• More soap is needed to form a lather, which increases costs. • It often leads to deposits (called scale) forming in heating systems and appliances like kettles. This reduces their efficiency.

Types of Hard Water

There are two types of hard water: **permanent** and **temporary**. Permanent hard water always contains a calcium salt (usually calcium sulfate) that is not removed by boiling, whereas temporary hard water can be softened by boiling.

HT Temporary hard water also contains a calcium salt (called calcium hydrogencarbonate). On heating the hydrogencarbonate (HCO_3^-) ions decompose to form carbonate ions (CO_3^{2-}). These carbonate ions react with magnesium or calcium ions to form precipitates of magnesium or calcium carbonate, removing the ions from the solution so the water is no longer hard. The precipitate is known as scale or limescale.

The hardness of water can be measured by titration with soap solution. Soap solution is added to a sample of hard water and the amount required to achieve a permanent lather is measured. The more soap solution needed to achieve a permanent lather, the harder the water. (See page 91 for a diagram of titration apparatus.)

Distinguishing Between Types of Hard Water

Soap solution is added to separate samples of temporary and permanent hard water until a permanent lather is achieved for a set time period (e.g. 10 seconds). The same amount of each sample of water is then boiled, and soap solution is added until a permanent lather is achieved for the same time period as the first test. The temporary hard water should require less soap solution after it has been boiled. This is because boiling softens temporary hard water.

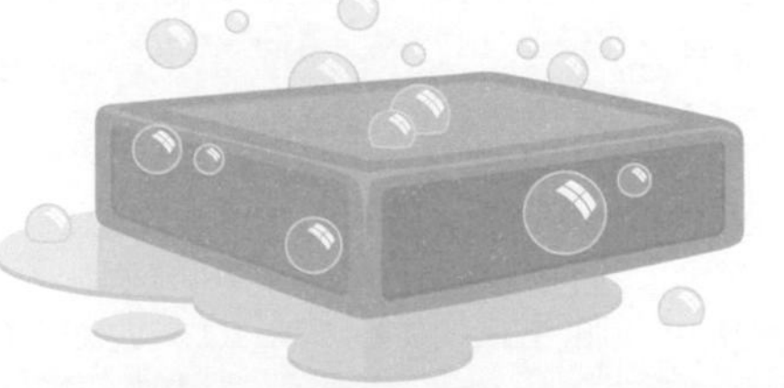

Removing Hardness

To make hard water (including permanent hard water) soft, we have to remove the dissolved calcium and magnesium ions it contains. This can be done in two ways:

- Add sodium carbonate solution (washing soda) to it. The carbonate ions react with the calcium and magnesium ions to form calcium carbonate and magnesium carbonate respectively, which precipitate out of solution because they are both insoluble. For example:

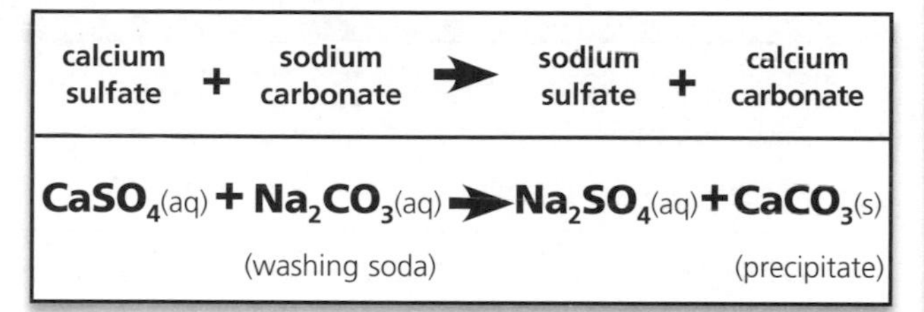

- Pass the hard water through an ion-exchange column. The column contains a special resin which supplies hydrogen ions, $H^+(aq)$ or sodium ions, $Na^+(aq)$. As the hard water passes through the resin, the calcium and magnesium ions in it are replaced by hydrogen or sodium ions from the resin. The calcium and magnesium ions consequently remain in the resin. The resin has to be replaced when it 'runs out' of hydrogen or sodium ions.

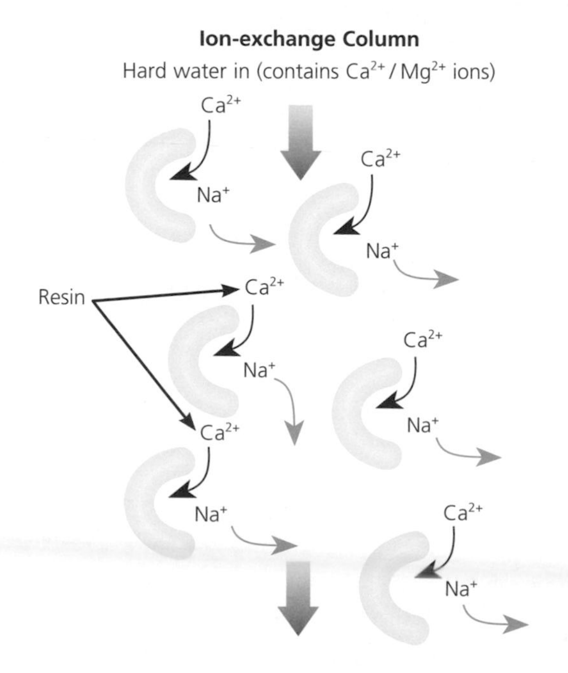

Purifying Water

Water is essential for life. In order to produce water that is of good quality and safe to drink, an appropriate water source is chosen. The water is passed through a filter bed to remove any solid particles, and chlorine gas or ozone is then added to kill any harmful microorganisms.

It is essential to ensure that the levels of dissolved salts in the water are sufficiently low. More dissolved substances can be removed from tap water to improve both the quality and taste by passing the water through a filter containing carbon, silver and ion-exchange resins.

Any water can be distilled to produce pure water, which contains no dissolved substances. This process involves boiling water to produce steam and then condensing the steam by cooling it to produce pure liquid water. This process uses a great deal of energy and is therefore very expensive.

Some substances are added to drinking water in order to benefit our health. Chlorine is added to drinking water to reduce the amount of microbes. Fluoride is sometimes added to improve dental health.

You need to be able to:

- **evaluate the use of commercial water softeners**
- **consider and evaluate the environmental, social and economic aspects of water quality and hardness**
- **consider the advantages and disadvantages of adding chlorine and fluoride to drinking water.**

Water Quality

It is important that drinking water is safe for us to use. In order to make sure water is safe to drink, solids are removed and chemicals are often added to kill bacteria that could cause disease.

Other factors need to be taken into consideration when looking at the quality of the water we use, e.g. the amount of nitrate. Fertilisers can run into the water supplies when it rains and excess levels of nitrates in drinking water can lead to health problems, particularly for young children and babies.

Some areas of the country have hard water, and some have softer water. Hardness of water can be removed, although there are advantages and disadvantages to removing the hardness.

Advantages of Hard Water
Contains calcium ions needed for healthy bones and teeth.
May help to reduce heart disease.
Some people prefer the taste compared with soft water.
Limescale build-up in pipes can prevent corrosion of pipes and stop some poisonous metal salts (e.g. from lead pipes) from dissolving in the water.
Disadvantages of Hard Water
Causes limescale to build up in water pipes, in heating systems/boilers and on kettle elements and increases fuel costs.
Build-up of limescale can cause pipes to become blocked.
More soap is required for washing.
Scum is formed when soap reacts with the calcium and magnesium ions present in hard water.

Adding Ions to Drinking Water

Fluoride ions may be added to drinking water because they are thought to strengthen tooth enamel and provide some protection against tooth decay. Despite the benefits, many people are opposed to adding fluoride ions to drinking water. They may believe that fluoride ions added are toxic or cause disease. Critics argue that as fluoride is often present in products such as toothpaste and mouthwash, adding it to the water supply is unnecessary and can lead to an accumulation of too much fluoride in the body. Many studies have been conducted into the safety of adding fluoride to drinking water and, although some people may be sensitive to fluoride added to water, there does not seem to be a link between the addition of fluoride and problems caused to the general public. Too much fluoride can cause discolouration of teeth, which, while undesired, does not cause harm. Chlorine is added to drinking water to kill bacteria and to decrease the risk of waterborne diseases. However, some people fear that adding chlorine to the water supply could lead to health problems.

Softening Water

Along with boiling, hard water can be softened by adding sodium carbonate (washing soda) or using an ion-exchange column. Adding sodium carbonate leads to the formation of precipitates of calcium carbonate or magnesium carbonate forming. This can be used to remove permanent hardness from water.

Ion-exchange columns work by exchanging the calcium and magnesium ions for sodium ions as water is passed through the column. The column is usually a glass tube filled with a resin that the hard water passes through. The columns have to be regenerated regularly because they become saturated with calcium and magnesium ions. The deposited ions are replaced with sodium ions, by running salt water (containing dissolved sodium chloride) through the column. This allows the column to be used again.

Scale inhibitors are used to stop the formation of limescale rather than to soften water.

You need to be able to consider and evaluate the environmental, social and economic aspects of water quality and hardness.

Example

When is pure not pure?

Water is the most abundant substance on the Earth's surface and it is important that water is fit to drink because polluted water can cause illnesses. The water we drink cannot be described as pure because it contains dissolved minerals it has collected on its way under and over the ground to the reservoir where it is stored. The World Health Organisation (WHO), The Environment Agency and Water Quality Association all employ people who carefully monitor what happens to our drinking water. Many human activities and their by-products have the potential to pollute water. Large and small industrial enterprises, agriculture, horticulture, transport, animals and humans can all bring about water pollution.

Water scare over

There was an outbreak of cryptosporidium – a stomach illness that can be caught by drinking infected water. In Anglesey and Gwynedd, there were 231 reported cases of people catching the bug. As a result, everyone in the North Wales area has had to boil their water before consuming it to remove the risk of getting infected.

The infected water is likely to have come from the Llyn Cwellyn reservoir, but health officials have now given North Wales residents the all clear; there is no need to continue boiling their water.

However, what remains to be seen is the impact of this scare. The general consensus is that people are still wary of their water, and some say they've switched to drinking only bottled water.

What's in our bottles?

There are plenty of brands of bottled water on the market – still, sparkling and even flavoured water – and the demand for them does not seem to be slowing down. When the average cost of a household's daily water usage is 68p, why are so many people happy to spend more than that on half a litre of bottled water?

There are many reasons why people choose to pay more to drink bottled water rather than drink the much cheaper alternative that comes out of their taps. Many people cite the fresher taste as their main reason for drinking bottled water. This fresher taste is because bottled water does not contain chlorine, which is used in the treatment of tap water. Interestingly, the drinking water inspectorate has claimed that most people cannot tell the difference between a glass of chilled tap water and a glass of chilled bottled water.

Others say they only buy bottled water as a convenience when they are out and about, and they choose water because it is a healthier alternative to the fizzy and sugary drinks on sale. However, many people claim they buy bottled water when they are out for the day, rather than taking a bottle of their own tap water, because it is chilled, which makes it more refreshing.

Another contributing factor is health fears, such as the contamination of tap water supplies. Bottled water companies latch on to this fear and use words such as 'clear', 'pure' and 'natural' to reinforce the idea that their water is clean and good for you. A leading foundation has recently stated that most worries about tap water have been exaggerated and that there is no reason why people should choose bottled water over tap water as far as their health is concerned.

However, even with the ever-increasing trend towards consuming bottled water rather than tap water, we still have a long way to go to catch up with the continent. In France, Germany and Italy, 90% of adults drink bottled water. Interestingly, they drink bottled water because of the minerals it contains and not because they think it is purer than tap water. They think the minerals are beneficial to health.

C3.3 Calculating and explaining energy change

The energy involved in a chemical reaction can be measured experimentally or be calculated. If we know the amount of energy involved in a reaction it can help us to use resources efficiently and economically. To understand this, you need to know:

- how to measure and calculate the amount of energy released when substances burn
- why energy is required and released in chemical reactions
- how to represent energy released / required, activation energy and the use of catalysts on energy level diagrams.

Joules and Calories

When any chemical change takes place it is accompanied by an energy change. If heat energy is given out it is an **exothermic reaction**, when it is taken in it is an **endothermic reaction** (see p.64).

The unit of measurement for energy is the **joule** (**J**). It takes 4.2 joules of energy to heat up 1g of water by 1°C. This amount of energy is called **1 calorie**.

Calculating the Amount of Energy

The amount of heat energy released when a substance burns, e.g. food or fuel, or when substances react can be calculated using the formula:

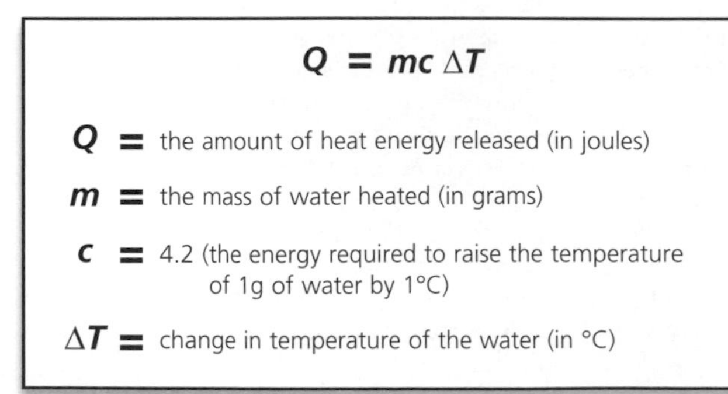

$$Q = mc\Delta T$$

Q = the amount of heat energy released (in joules)

m = the mass of water heated (in grams)

c = 4.2 (the energy required to raise the temperature of 1g of water by 1°C)

ΔT = change in temperature of the water (in °C)

Remember, if you need to convert between joules (J) and kilojoules (kJ) in your answer: 1000J = 1kJ.

This formula can also be used to calculate the amount of energy released when solids react / dissolve in water or in neutralisation reactions.

Measuring Energy

The method of measuring the amount of heat energy released by burning a substance is called **calorimetry** because the equipment used is called a calorimeter. Calorimeters are usually made from metal or glass.

To measure the temperature change that takes place when a fuel burns you need to do the following:

1. Place 100g of water in a calorimeter and measure the temperature of the water.
2. Find the mass (in grams) of the fuel to be burned.
3. Burn the fuel under the water in the calorimeter for a few minutes and record the temperature change of the water.
4. Weigh the fuel again to be able to calculate how much fuel has been used.

The amount of energy released in a chemical reaction in solution can be measured by mixing the reactants in an insulated container, which enables the temperature change to be measured before heat is lost to the surroundings. This method would be suitable for neutralisation reactions and the reaction of solids and water.

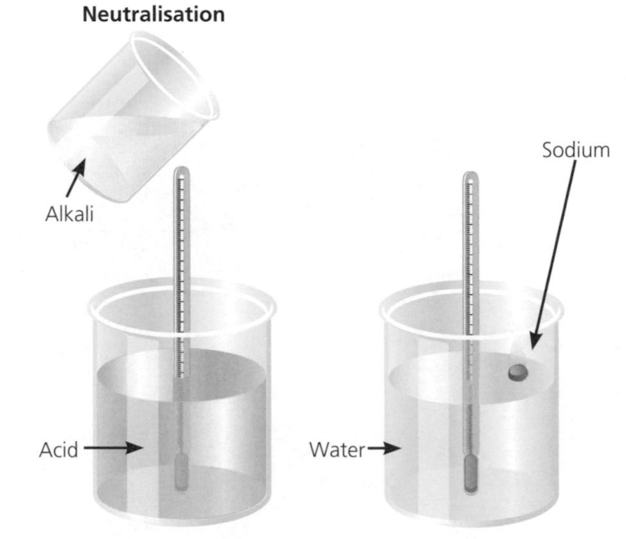

Temperature rise caused by adding the alkali to the acid

Temperature rise by adding the sodium to the water

Example

Calculate the energy per gram produced by burning methylated spirits (meths).

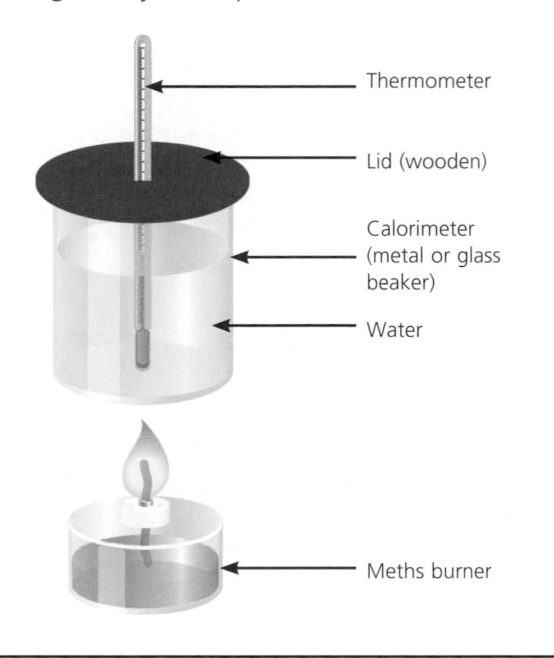

Temperature change	=	Highest temperature reached	–	Temperature at start

	Before	After	Difference
Mass of meths and burner	125.0g	122.6g	125.0 – 122.6 = 2.4g
Temperature of 100g of water	21°C	37°C	37 – 21 = 16°C

So, 2.4g of meths raises 100g of water by 16°C.

We know that 4.2 joules of energy will raise the temperature of 1g of water by 1°C, so we can use this to calculate how many joules it will take to heat 100g of water:

$Q = m \times c \times \Delta T$

$= 100 \times 4.2 \times 16 = 6720\text{J}$

So, now we know that 2.4g of meths produces 6720 joules of energy, we can work out how much energy 1g of meths will produce:

$\frac{6720}{2.4} = 2800\text{J} =$ **2.8kJ/g**

Making and Breaking Bonds

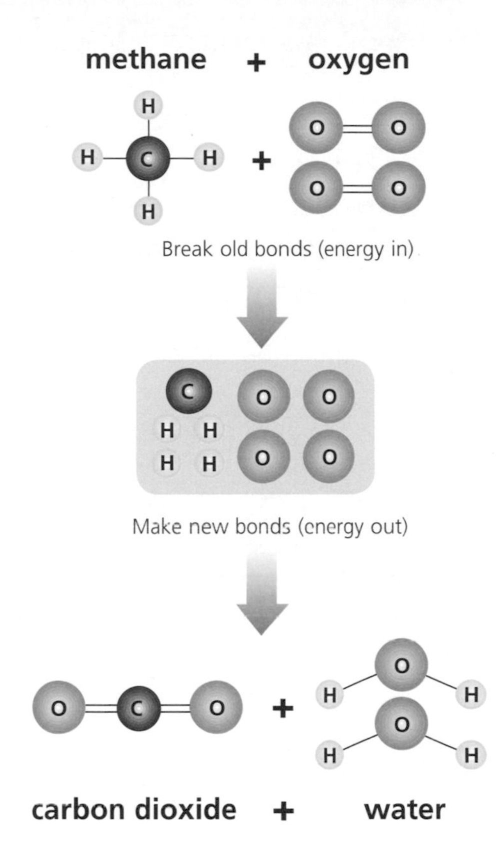

In a chemical reaction, new substances are produced. In order to do this, the bonds in the reactants must be broken and new bonds made to form the products.

Breaking a chemical bond requires energy – this must be an **endothermic** process.

When a new chemical bond is formed, energy is given out – this must be an **exothermic** process.

We can use this idea to find out if a chemical reaction is exothermic or endothermic overall (see p. 87).

HT **Endothermic**

If more energy is required to break old bonds than is released when the new bonds are formed, the reaction must be **endothermic**.

Exothermic

If more energy is released when the new bonds are formed than is needed to break the old bonds, the reaction must be **exothermic**.

Energy Level Diagrams

The energy changes in a chemical reaction can be illustrated using an energy level diagram.

Exothermic Processes

In an exothermic reaction, energy is given out. This means energy is being lost so the products have less energy than the reactants.

Exothermic reactions have a negative energy change.

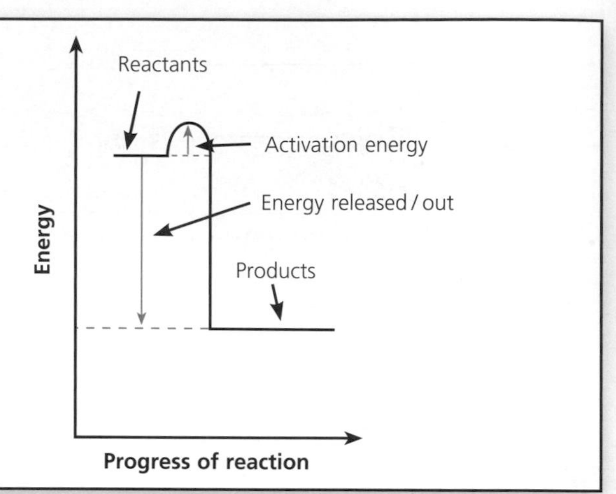

Endothermic Processes

In an endothermic reaction, energy is taken in. This means that energy is being gained, so the products have more energy than the reactants.

Endothermic reactions have a positive energy change.

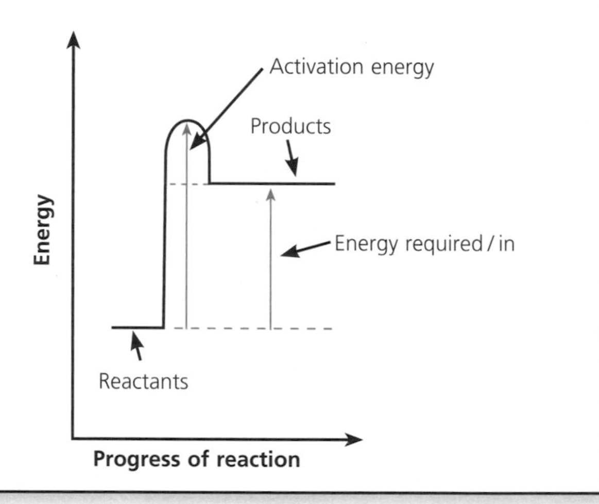

Activation Energy

The activation energy is the minimum energy needed for a reaction to occur, i.e. to break the old bonds (see above diagrams). We can show this on an energy level diagram too.

Catalysts

Catalysts speed up the rate of a chemical reaction by providing an alternative pathway for the reaction to take with lower activation energy. This also means that the reaction can be carried out at a lower temperature.

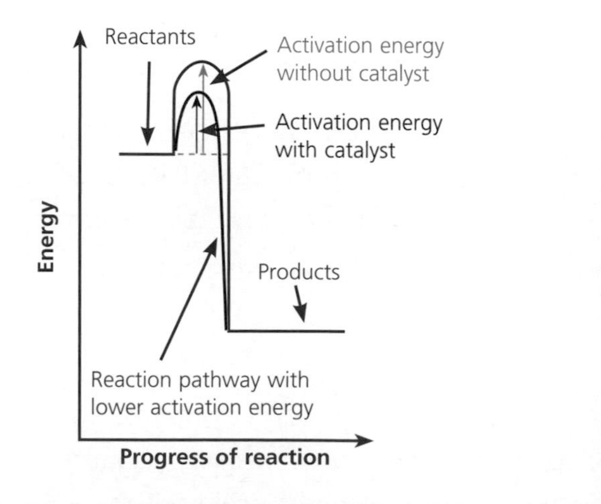

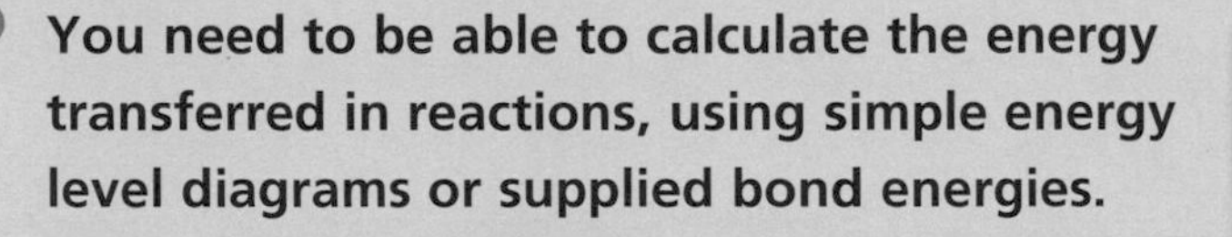

You need to be able to calculate the energy transferred in reactions, using simple energy level diagrams or supplied bond energies.

Supplied Bond Energies

We can find out whether a reaction is exothermic or endothermic by calculating the difference between the energy released when new bonds are made and the energy used to break bonds. For example, when methane burns, the products formed are carbon dioxide and water. The reaction can be represented by the equation below:

methane + oxygen ⟶ carbon dioxide + water
$CH_{4(g)} + 2O_{2(g)} \longrightarrow CO_{2(g)} + 2H_2O_{(l)}$

Bond energies needed for this are:

C–H is 412kJ/mol
O=O is 496kJ/mol
C=O is 805kJ/mol
H–O is 463kJ/mol

Energy used to break bonds is:

4 C–H = 4 × 412 = 1648kJ
2 O=O = 2 × 496 = 992kJ
Total = 1648kJ + 992kJ = **2640kJ**

Energy given out by making bonds:

2 C=O = 2 × 805 = 1610kJ
4 H–O = 4 × 463 = 1852kJ
Total = 1610kJ + 1852kJ = **3462kJ**

Energy change (ΔH) = Energy used to break bonds – Energy given out by making bonds
= 2640kJ – 3462kJ = **–822kJ**

The reaction is exothermic (energy is given out) because the energy from making the bonds in the products is more than the energy needed for breaking the bonds in the reactants.

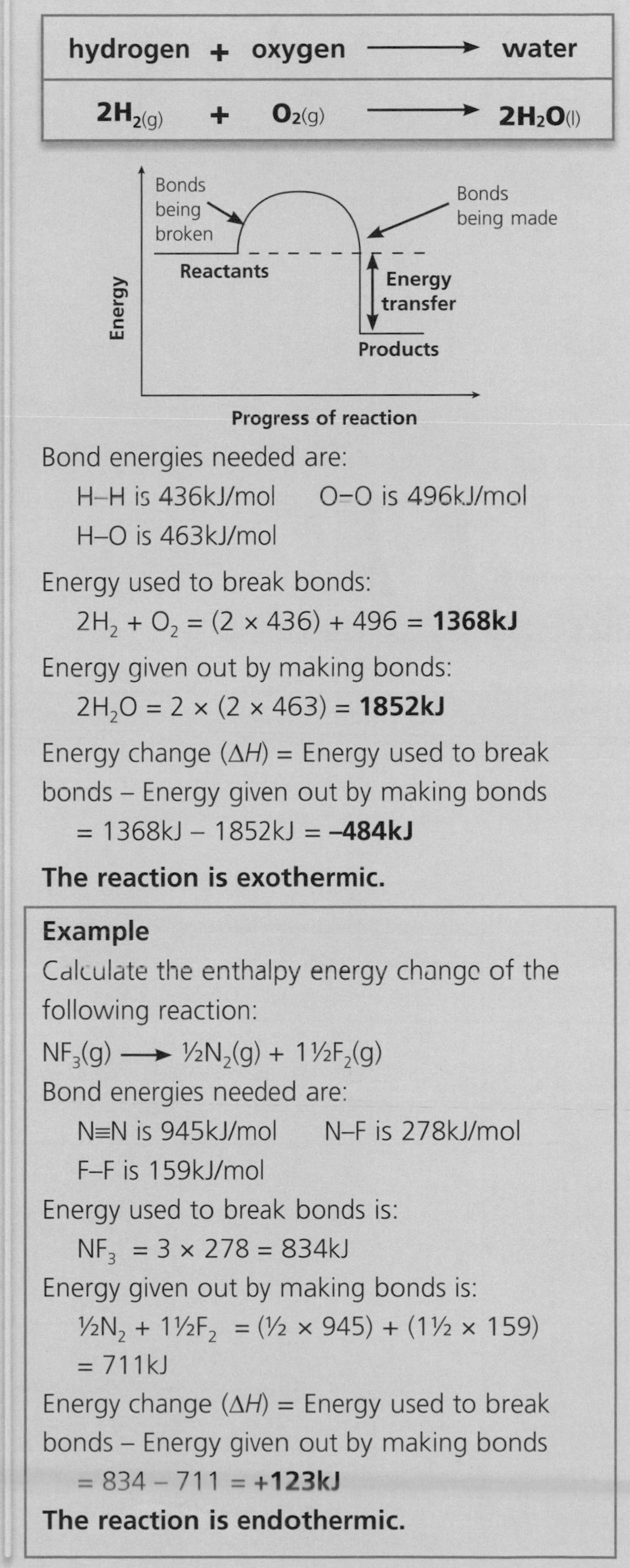

Simple Energy Level Diagrams

Hydrogen and oxygen are burned to produce water.

hydrogen + oxygen ⟶ water
$2H_{2(g)} + O_{2(g)} \longrightarrow 2H_2O_{(l)}$

Bond energies needed are:

H–H is 436kJ/mol O=O is 496kJ/mol
H–O is 463kJ/mol

Energy used to break bonds:

$2H_2 + O_2$ = (2 × 436) + 496 = **1368kJ**

Energy given out by making bonds:

$2H_2O$ = 2 × (2 × 463) = **1852kJ**

Energy change (ΔH) = Energy used to break bonds – Energy given out by making bonds
= 1368kJ – 1852kJ = **–484kJ**

The reaction is exothermic.

Example

Calculate the enthalpy energy change of the following reaction:

$NF_3(g) \longrightarrow \frac{1}{2}N_2(g) + 1\frac{1}{2}F_2(g)$

Bond energies needed are:

N≡N is 945kJ/mol N–F is 278kJ/mol
F–F is 159kJ/mol

Energy used to break bonds is:

NF_3 = 3 × 278 = 834kJ

Energy given out by making bonds is:

$\frac{1}{2}N_2 + 1\frac{1}{2}F_2$ = (½ × 945) + (1½ × 159)
= 711kJ

Energy change (ΔH) = Energy used to break bonds – Energy given out by making bonds
= 834 – 711 = **+123kJ**

The reaction is endothermic.

You need to be able to compare the energy produced by different fuels and foods and consider the social, economic and environmental consequences of using fuels.
Example

FACTSHEET

Most fossil-fuelled power stations burn coal or natural gas. The main products from burning coal in a power station are carbon dioxide (CO_2) and smoke. The increase of carbon dioxide levels in the atmosphere is believed to be responsible for the greenhouse effect which could result in global warming. Smoke contains microscopic particles of carbon, called particulates. Although they are small, they are much bigger than atoms and molecules and each particle contains billions of carbon atoms. It is claimed that they can make asthma and lung infections worse.

Over the last two and a half years, Drax Power Station has been developing the use of biomass as a fuel to use in addition to coal in order to try to reduce emissions of carbon dioxide.

Biomass has replaced 2.5% of coal at Drax and this has reduced emissions of carbon dioxide by around 0.5 million tonnes a year. Further modifications at Drax could save a remarkable 4.4 million tonnes a year.

The National Farmers' Union (NFU) president has said that British farmers have a vested interest in reducing the impact of climate change.

Out of all energy crops, willow represents the least intensive. It can be grown and harvested every 3 years to produce wood chip, and a willow plantation can remain viable for up to 20 years.

Fuel	Advantages	Disadvantages
Biomass (e.g. willow)	• Less CO_2 emissions. • Renewable energy source, so it is sustainable. • Burning wood does not contribute as much to global warming as burning fossil fuels. • Carbon neutral – plants take in CO_2 during their lifetime and this is released when they are burned. Therefore, they are classed as carbon neutral.	• Produces SO_2 which causes acid rain. • Large area needed to grow trees, which could be used for another purpose. • Not very efficient if small plants are used. • Not as much energy (heat) produced as there is from coal. • Adverse effects on health, particularly for asthma sufferers.
Fossil fuel (e.g. coal, gas)	• Large amounts of energy can be produced very cheaply. • Gas power stations are very efficient.	• Non-renewable energy source (will run out). • Produces CO_2 which contributes to the greenhouse effect. • Produces SO_2, which causes acid rain. • Produces particulates and pollution.
Hydrogen	• Does not produce pollutants when burned (the only product is water) $2H_2 + O_2 \rightarrow 2H_2O$ • Can be used in some combustion engines as a fuel or in fuel cells to produce electricity to power vehicles.	• Difficult to store and transport hydrogen. • Most hydrogen currently made from non-renewable resources, e.g. reacting steam with coal or methane. • Some hydrogen is produced by electrolysis of water – requires large amounts of electricity that is currently generated from non-renewable resources.

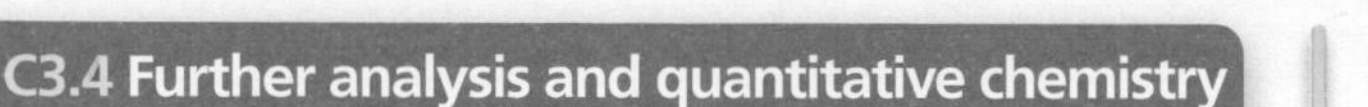

C3.4 Further analysis and quantitative chemistry

Chemical tests can identify elements and compounds even when only small samples are available. To understand this, you need to know:

- how and why flame tests are used
- how carbonates react with acids
- how to identify metal ions using sodium hydroxide solution
- how to identify some non-metal ions
- how titrations can be used.

A range of chemical tests can be carried out to detect and identify elements and compounds.

Flame Tests

Flame tests are used to identify the metal ions present in a compound. Lithium, sodium, potassium, calcium and barium compounds can be recognised by the distinctive colours they produce in a flame test.

1. A piece of nichrome (a nickel-chromium alloy) wire is dipped in concentrated hydrochloric acid to clean it.

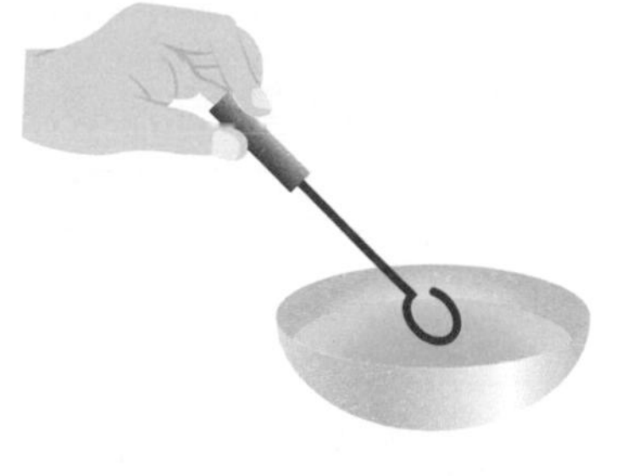

2. It is dipped in the compound …

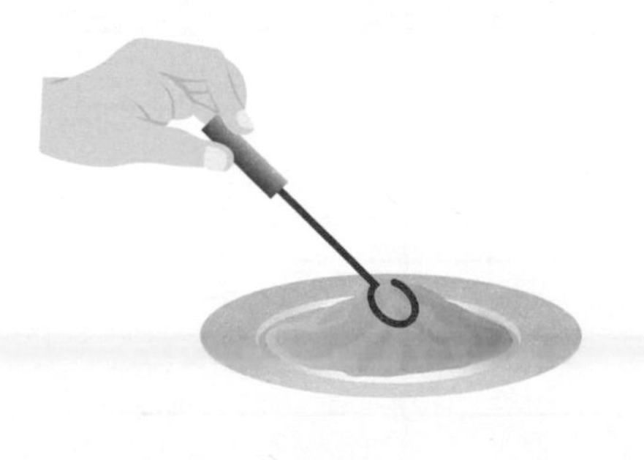

3. … and then put into a Bunsen flame to give the following distinctive colours:

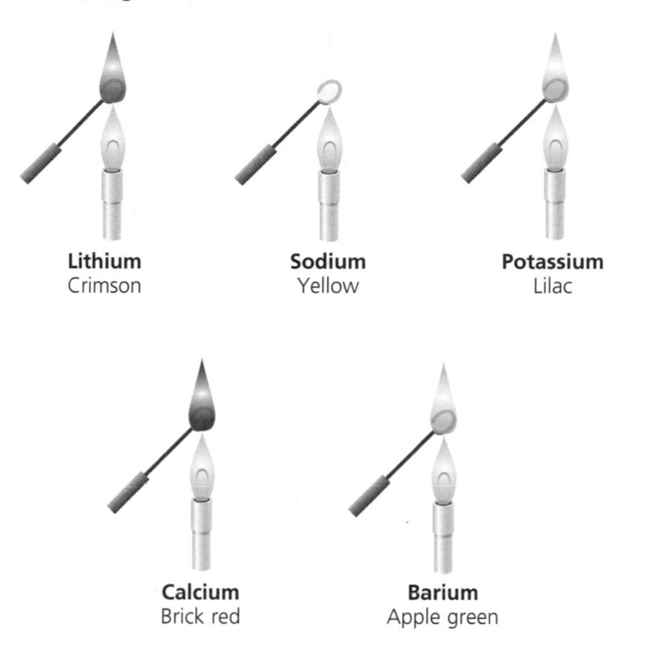

Reaction of Carbonates with Dilute Acid

Carbonates react with dilute acids to form carbon dioxide gas (and a salt and water). For example, if we add calcium carbonate to dilute hydrochloric acid then the carbonate will 'fizz' as it reacts with the acid, giving off carbon dioxide.

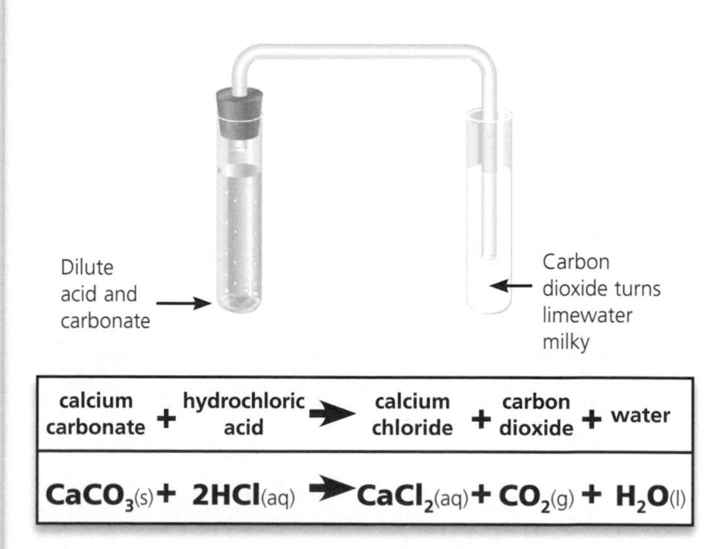

calcium carbonate + hydrochloric acid → calcium chloride + carbon dioxide + water

$$CaCO_{3(s)} + 2HCl_{(aq)} \rightarrow CaCl_{2(aq)} + CO_{2(g)} + H_2O_{(l)}$$

When carbon dioxide is bubbled through limewater, the limewater turns cloudy. This is due to the white precipitate of calcium carbonate that is formed when limewater and carbon dioxide react.

Metal Ions

Metal compounds in solution contain metal ions. Some of these form precipitates – i.e. insoluble solids – that come out of solution when sodium hydroxide solution is added to them. In the example below, a white precipitate of calcium hydroxide is formed (as well as sodium chloride solution). We can see how this precipitate is formed by considering the ions involved:

$$Ca^{2+}(aq) + 2OH^{-}(aq) \longrightarrow Ca(OH)_2(s)$$

$Ca^{2+}(aq)$ from the calcium chloride; $2OH^{-}(aq)$ from the sodium hydroxide

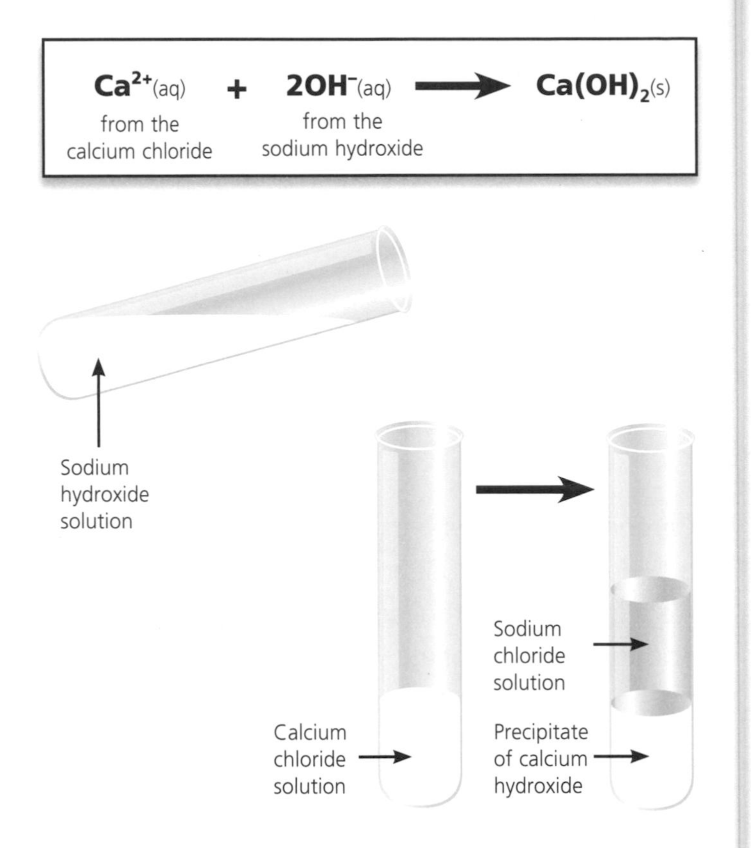

The table below shows the colour of the precipitates formed when certain metal ions are mixed with sodium hydroxide solution. Only the aluminium hydroxide precipitate dissolves in excess sodium hydroxide solution.

Sulfate Ions

To identify the presence of sulfate ions, add barium chloride solution and dilute hydrochloric acid to the suspected sulfate solution. A white precipitate of barium sulfate will be produced if a sulfate is present.

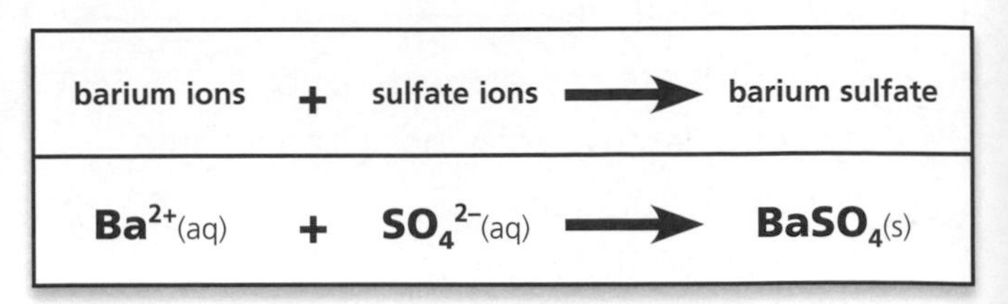

barium ions + sulfate ions ⟶ barium sulfate

$$Ba^{2+}(aq) + SO_4^{2-}(aq) \longrightarrow BaSO_4(s)$$

Halide Ions (Chloride, Bromide and Iodide)

To identify the presence of a chloride, bromide or iodide ion, add silver nitrate solution and dilute nitric acid to the suspected halide solution. A white precipitate will form if silver chloride is present, a cream precipitate for silver bromide, and a yellow precipitate for silver iodide.

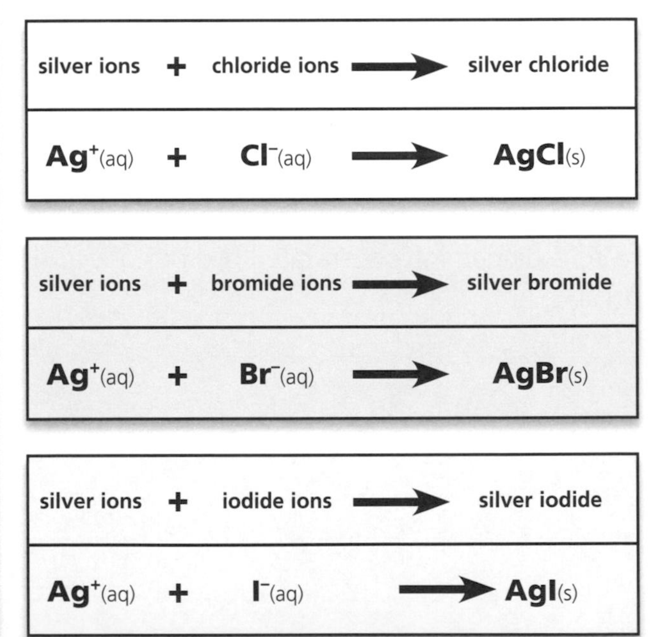

silver ions + chloride ions ⟶ silver chloride

$$Ag^{+}(aq) + Cl^{-}(aq) \longrightarrow AgCl(s)$$

silver ions + bromide ions ⟶ silver bromide

$$Ag^{+}(aq) + Br^{-}(aq) \longrightarrow AgBr(s)$$

silver ions + iodide ions ⟶ silver iodide

$$Ag^{+}(aq) + I^{-}(aq) \longrightarrow AgI(s)$$

Metal Ion + Sodium Hydroxide		Precipitate Formed	Colour of Precipitate
aluminium	$Al^{3+}_{(aq)}$ + sodium hydroxide	aluminium hydroxide	white
calcium	$Ca^{2+}_{(aq)}$ + sodium hydroxide	calcium hydroxide	white
magnesium	$Mg^{2+}_{(aq)}$ + sodium hydroxide	magnesium hydroxide	white
copper	$Cu^{2+}_{(aq)}$ + sodium hydroxide	copper hydroxide	blue
iron(II)	$Fe^{2+}_{(aq)}$ + sodium hydroxide	iron(II) hydroxide	green
iron(III)	$Fe^{3+}_{(aq)}$ + sodium hydroxide	iron(III) hydroxide	brown

Titration

Titration is an accurate technique that can be used to find out how much of an acid is needed to neutralise an alkali. If the concentration of the acid (mol/dm^3) is known, the concentration of the alkali can be worked out, or vice versa.

A pipette, which has been carefully washed and rinsed with the alkali, is used to measure out a known and accurate volume of the alkali. The alkali is then placed in a clean and dry conical flask and a suitable indicator (e.g. phenolphthalein) is added. Next, acid is placed in a burette, which has been carefully washed and rinsed with the acid. An initial reading of the volume of acid in the burette is made.

The acid is carefully added to the alkali until the indicator changes colour (to show neutrality). This is called the **end point**. A final reading is taken of the volume of acid remaining in the burette. You can calculate the volume of acid added.

The method can be repeated to check results; and can then be performed without an indicator in order to obtain the salt if required.

You must use a suitable indicator to find the volumes of different strength acids and alkalis in neutralisation reactions. For example, when a strong acid and a strong alkali react any suitable acid-base indicator (e.g. phenolphthalein, methyl orange) can be used.

When neutralisation takes place, the hydrogen ions (H^+) from the acid join with the hydroxide ions (OH^-) from the alkali to form water, which is neutral.

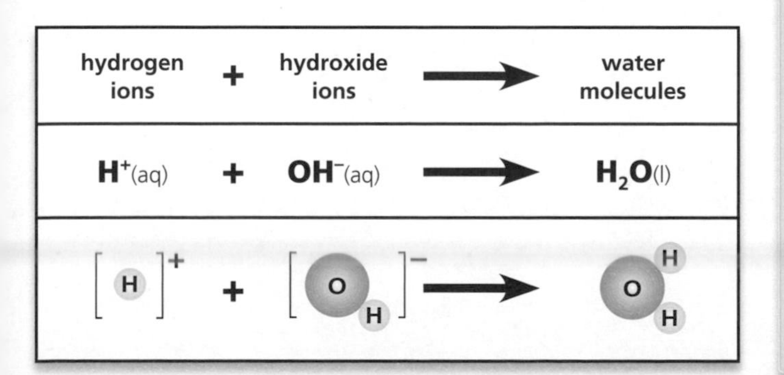

hydrogen ions	+	hydroxide ions	⟶	water molecules
$H^+(aq)$	+	$OH^-(aq)$	⟶	$H_2O(l)$
$[H]^+$	+	$[OH]^-$	⟶	H_2O

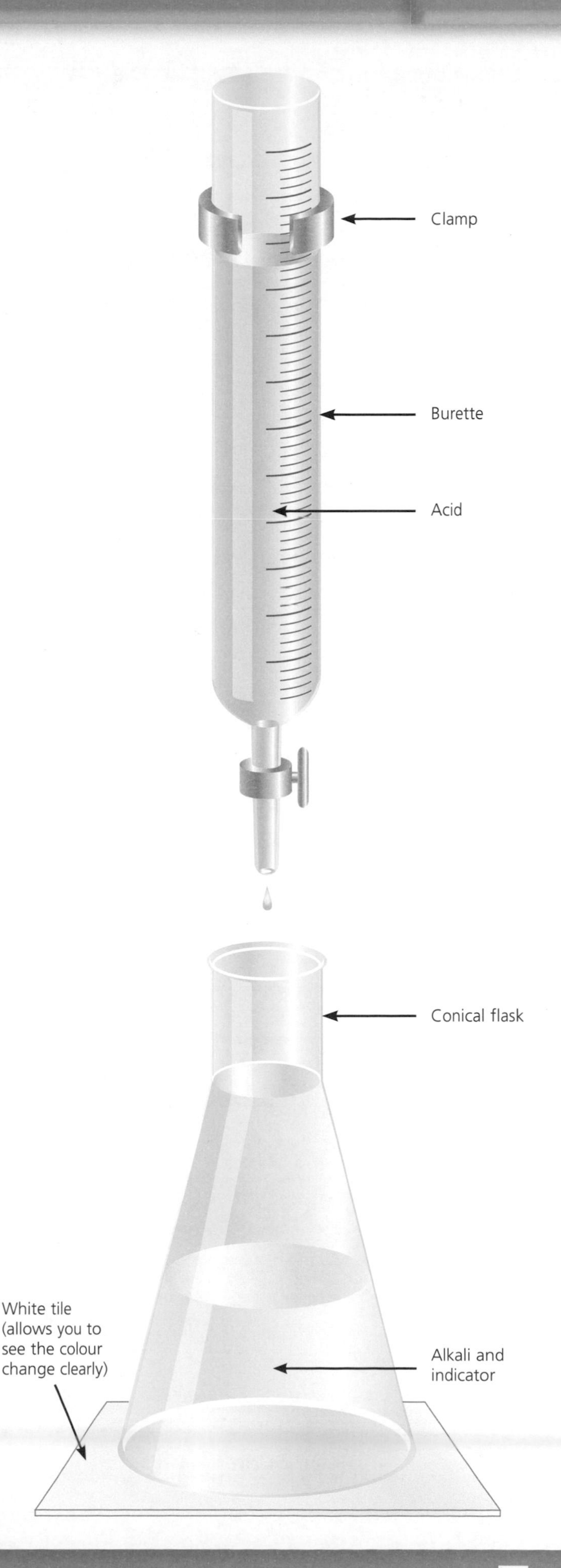

HT Titration (cont)

Titration can be used to find the concentration of an acid or alkali providing you know either:

- the relative volumes of acid and alkali used
- the concentration of the other acid or alkali.

You will need to break down the calculation into three stages.

1. Write down a balanced equation for the reaction taking place to determine the ratio of moles of acid to alkali involved.
2. Calculate the number of moles in the solution of known volume and concentration. You will know the number of moles in the other solution from stage 1.
3. Calculate the concentration of the other solution using the formula:

$$\text{Concentration of solution (mol/dm}^3\text{)} = \frac{\textbf{Number of moles of solute} \text{ (mol)}}{\textbf{Volume of solution} \text{ (dm}^3\text{)}}$$

Example 1

A titration is carried out and 0.04dm^3 of hydrochloric acid neutralises 0.08dm^3 of sodium hydroxide of concentration 1mol/dm^3. Calculate the concentration of the hydrochloric acid.

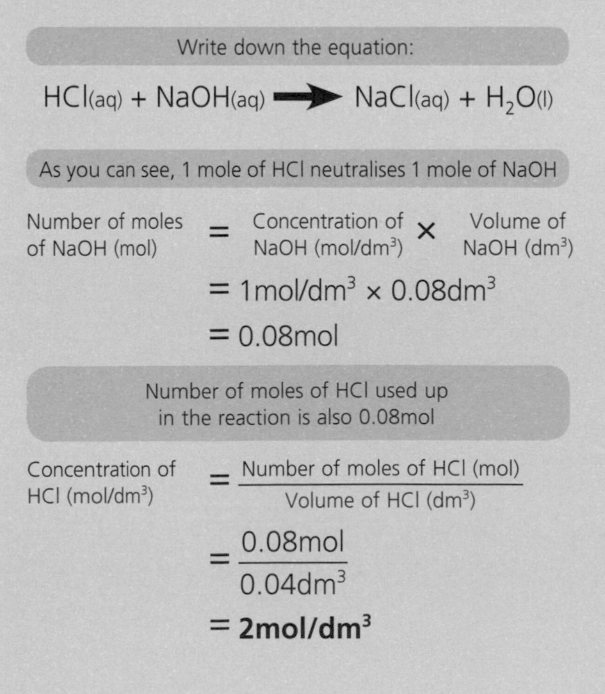

Write down the equation:

$$HCl_{(aq)} + NaOH_{(aq)} \longrightarrow NaCl_{(aq)} + H_2O_{(l)}$$

As you can see, 1 mole of HCl neutralises 1 mole of NaOH

$$\text{Number of moles of NaOH (mol)} = \text{Concentration of NaOH (mol/dm}^3\text{)} \times \text{Volume of NaOH (dm}^3\text{)}$$

$$= 1\text{mol/dm}^3 \times 0.08\text{dm}^3$$

$$= 0.08\text{mol}$$

Number of moles of HCl used up in the reaction is also 0.08mol

$$\text{Concentration of HCl (mol/dm}^3\text{)} = \frac{\text{Number of moles of HCl (mol)}}{\text{Volume of HCl (dm}^3\text{)}}$$

$$= \frac{0.08\text{mol}}{0.04\text{dm}^3}$$

$$= \mathbf{2mol/dm^3}$$

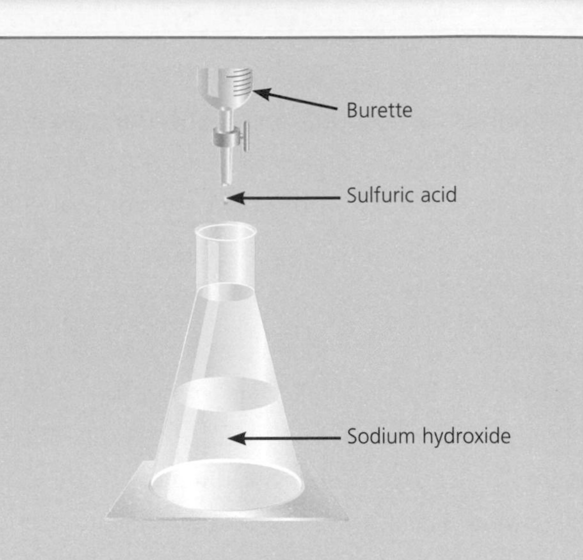

Example 2

A titration is carried out and 0.035dm^3 of sulfuric acid of concentration 0.6mol/dm^3 neutralises 0.14dm^3 of sodium hydroxide. Calculate the concentration of sodium hydroxide in g/dm^3.

Write down the equation:

$$H_2SO_{4}(aq) + 2NaOH(aq) \longrightarrow Na_2SO_{4}(aq) + 2H_2O(l)$$

This time, 1 mole of H_2SO_4 neutralises 2 moles of NaOH.

$$\text{Number of moles of } H_2SO_4 \text{ (mol)} = \text{Concentration of } H_2SO_4 \text{ (mol/dm}^3\text{)} \times \text{Volume of } H_2SO_4 \text{ (dm}^3\text{)}$$

$$= 0.6\text{mol/dm}^3 \times 0.035\text{dm}^3$$

$$= 0.021\text{mol}$$

Number of moles of NaOH used up in the reaction is $2 \times 0.021 = 0.042\text{mol}$

$$\text{Concentration of NaOH (mol/dm}^3\text{)} = \frac{\text{Number of moles of NaOH (mol)}}{\text{Volume of NaOH (dm}^3\text{)}}$$

$$= \frac{0.042\text{mol}}{0.14\text{dm}^3}$$

$$= 0.3\text{mol/dm}^3$$

To convert concentration in mol dm^{-3} to concentration in g dm^{-3} we use the following formula:

$$\mathbf{g/dm^3 = mol/dm^3 \times} \boldsymbol{M_r}$$

(where M_r = relative formula mass of compound)

$$\text{Concentration of NaOH (g/dm}^3\text{)} = \text{Concentration of NaOH (mol/dm}^3\text{)} \times M_r \text{ NaOH}$$

$$= 0.3 \times 40$$

$$= \mathbf{12\ g/dm^3}$$

C3.5 The production of ammonia

Ammonia is made using the Haber process. Energy requirements need to be carefully considered for the industrial process to ensure the maximum yield possible is achieved. To understand this, you need to know:

- the raw materials and conditions required for the Haber process
- how equilibrium can be reached in a reversible reaction
- what effect changing temperature and pressure can have on an equilibrium reaction
- the yield depends on the conditions of the reaction.

The Production of Ammonia

Ammonia (NH_3) is made by the Haber process. The raw materials for this process are:

- nitrogen – obtained from the fractional distillation of air
- hydrogen – obtained from the reaction of natural gas and steam.

The purified nitrogen and hydrogen are passed over an iron catalyst at high temperature (about 450°C) and pressure (about 200 atmospheres). Some of the nitrogen and hydrogen will react to form ammonia, but the reaction is reversible so ammonia breaks down again into nitrogen and hydrogen:

nitrogen	+	hydrogen	⇌	ammonia
$N_2(g)$	+	$3H_2(g)$	⇌	$2NH_3(g)$

The ammonia gas liquefies as it cools and can be tapped off. Although the reaction conditions are chosen to produce a reasonable yield of ammonia, only some of the nitrogen and hydrogen will react. Therefore, any unreacted nitrogen and hydrogen is recycled.

Sustainable Development

It is important for sustainable development and economic reasons to minimise energy use and wastage in industrial processes. Non-vigorous conditions (i.e. reducing the temperature) mean that less energy is used and less is released into the environment, but the reaction is slower.

HT

Reversible Reactions in Closed Systems

When a reversible reaction occurs in a closed system (where no reactants are added and no products are removed) then an equilibrium is achieved in which the reactions occur at exactly the same rate in both directions. The relative amounts of all the reacting substances at equilibrium depend on the conditions of the reaction. If we take the reaction:

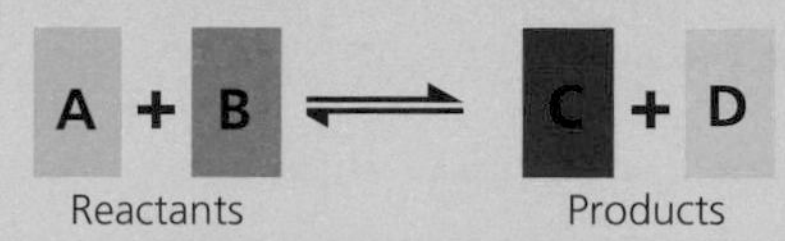

If the forward reaction (the reaction that produces the products C and D) is endothermic then:

If the temperature is increased, the yield of products is increased.

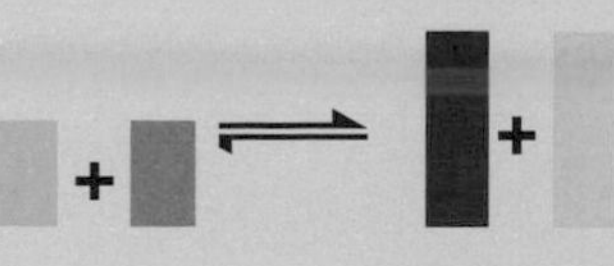

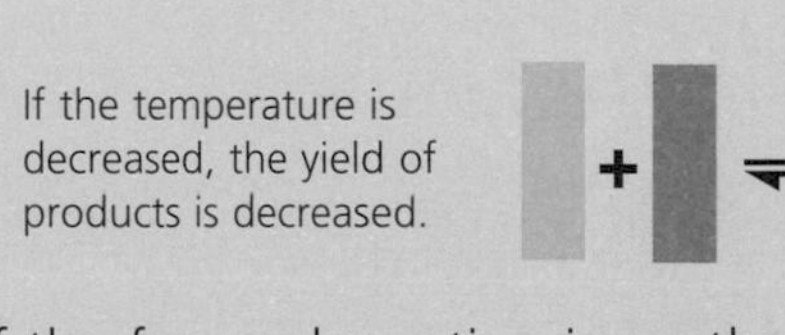

If the forward reaction is exothermic then:

If the temperature is increased, the yield of products is decreased.

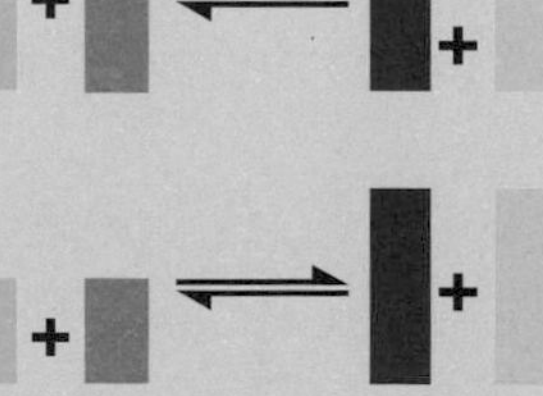

When the temperature of a system at equilibrium is increased, the endothermic reaction is favoured because this will oppose the change by 'using up the heat energy' or decreasing the temperature. Endothermic reactions require heat energy.

Reversible Reactions in Closed Systems (cont)

When the temperature of a system at equilibrium is decreased, the exothermic reaction is favoured because this will oppose the change by increasing the temperature. Exothermic reactions release heat energy.

In an equilibrium reaction involving gases, an increase in pressure will favour the reaction that produces the smallest number of molecules of gas. A decrease in pressure will favour the reaction that produces the biggest number of molecules of gas. For example:

An increase in pressure will favour the forward reaction (3 gas molecules on the left-hand side and 2 gas molecules on the right-hand side) so more C will be produced.

A decrease in pressure will favour the backward reaction (more gas molecules on the left than on the right) so more A and B_2 would be produced.

Even though a reversible reaction may not go to completion, it may still be used efficiently in an industrial process, e.g. the Haber process. Factors affecting equilibrium, together with reaction rates, are important when determining the optimum conditions in industrial processes.

Effect of Varying Conditions on Reversible Reactions

The manufacture of ammonia is a reversible reaction, involving energy transfers associated with the breaking and formation of chemical bonds.

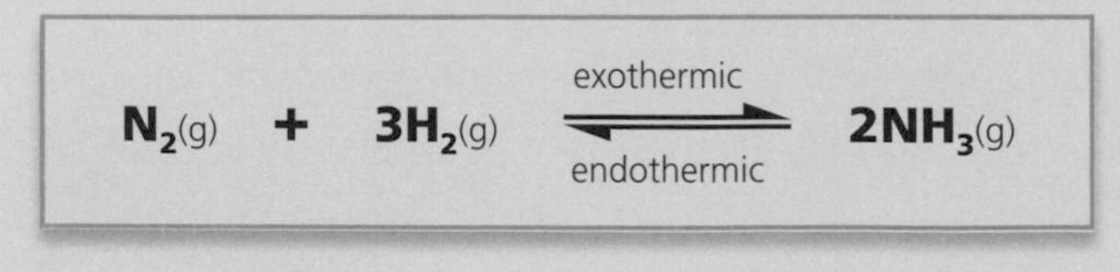

Less energy is needed to break the bonds between the nitrogen and hydrogen molecules than is released in the formation of the ammonia molecules.

Effect of Temperature

At low temperature, the production of ammonia (the forward reaction), which is an exothermic reaction, is favoured, i.e. the yield of ammonia is increased.

Increasing the temperature increases the rate of reaction equally in both directions, therefore high temperatures make ammonia form faster, but also break down faster.

Effect of Pressure

Increasing the pressure favours the smaller volume (side with the smallest number of molecules of gas). Therefore, using a high pressure favours the production of ammonia (the forward reaction), since there are four molecules of gas on the left-hand side and only two molecules of gas on the right-hand side. Increasing pressure, therefore, increases the yield of ammonia. Increasing the pressure also increases the rate of reaction.

A Compromise Solution

Altering the temperature and pressure can have a big impact on the production of ammonia in the Haber process. The conditions have to be chosen very carefully to be economically viable and to make sure they can meet demand.

The formation of ammonia is exothermic, so a low temperature increases the yield, but the reaction is very slow. A high temperature makes the reaction faster, but produces a lower yield. So a compromise is reached (450°C).

The volume of ammonia produced is less than the total volume of the reactants (nitrogen and hydrogen) so a high pressure favours the production of ammonia – but this is very expensive. A low pressure is more affordable and safer, but this produces a lower yield. So yet again a compromise is reached (200 atmospheres).

You need to be able to evaluate the conditions used in industrial processes in terms of energy requirements.

Chemical reactions involve energy transfers. Many chemical reactions involve the release of energy and some reactions take in energy. In industrial processes, energy requirements and emissions need to be considered both for economic reasons and for sustainable development. It is important that energy is not wasted in industrial processes. Non-vigorous conditions means less energy is used and less is lost into the environment (i.e. carried out at a lower temperature).

The conditions used in any industrial process must be considered in terms of the yield of the product. There might need to be a compromise between maximum yield and speed of reaction.

It could be more economical to wait longer for the sake of a higher yield.

Using a Catalyst

Adding a catalyst lowers the activation energy for a reaction, so lower temperatures can be used. This usually results in a saving in energy costs.

Changing Pressure or Temperature

Raising the temperature of a chemical reaction increases the energy of the particles involved and they move faster. If they move faster, more successful collisions between the particles take place and this means the reaction goes faster. Sometimes, however, the higher temperature can encourage some of the products made to break up and so this is not favourable for the forward reaction. A compromise temperature is chosen for the process.

Increasing the pressure also increases the rate of reaction because the particles have less room to spread out in, so they collide more often.

When evaluating the methods used in industry, we need to take into account the amount of energy needed for a reaction, the percentage yield that each method makes, the cost and the impact on the environment.

Using a Catalyst

Advantages
- Reduces the activation energy needed for a reaction.
- Speeds up a reaction.
- Reduces the energy cost of a reaction.

Disadvantages
- Different reactions need different catalysts.
- Catalyst needs to be removed from the products and cleaned regularly to prevent it from becoming poisoned.
- Purchasing catalysts can be costly.

Increasing Pressure or Temperature

Advantages
- Increases the rate of reaction.
- May get a higher percentage yield.

Disadvantages
- The reaction and the plant machinery would cost more.
- The percentage yield will decrease if the reaction is exothermic.

C3.6 Alcohols, carboxylic acids and esters

Alcohols and carboxylic acids are organic compounds with many different uses. To understand this, you need to know:

- the names and formula for the first three alcohols and carboxylic acids
- the properties and reactions of alcohols and carboxylic acids
- how to recognise and form esters
- the uses of alcohols, carboxylic acids and esters.

Alcohols

Alcohols are organic compounds containing carbon, hydrogen and oxygen. They contain the **functional group** –OH and have the general formula: $C_nH_{2n+1}OH$.

Alcohols are members of a **homologous series**. The first three members of that series are shown in the table:

Name	Formula	Displayed Formula
Methanol	CH_3OH	H–C(H)(H)–O–H
Ethanol	C_2H_5OH	H–C(H)(H)–C(H)(H)–O–H
Propanol	C_3H_7OH	H–C(H)(H)–C(H)(H)–C(H)(H)–O–H

All alcohols have names ending in -ol.

Small-molecule alcohols are soluble in water and will dissolve to form a neutral solution.

Alcohols are very useful compounds and are often used as fuels or solvents. Ethanol is the main alcohol in alcoholic drinks.

Reactions of Alcohols

Alcohols undergo the following reactions.

- Reaction with oxygen in the air (combustion reaction):

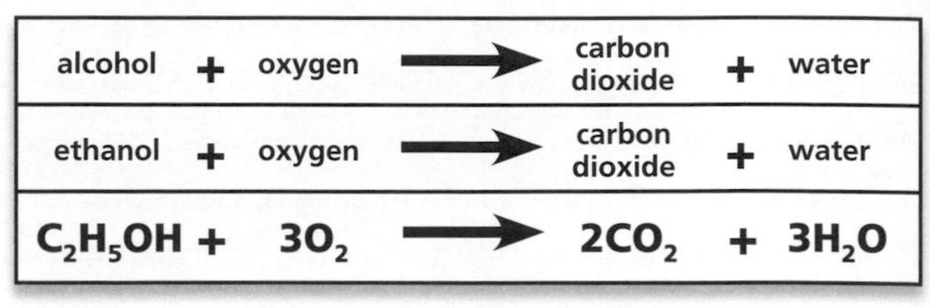

alcohol	+	oxygen	→	carbon dioxide	+	water
ethanol	+	oxygen	→	carbon dioxide	+	water
C_2H_5OH	+	$3O_2$	→	$2CO_2$	+	$3H_2O$

- Reaction with sodium to form hydrogen gas:

alcohol	+	sodium	→	sodium alkoxide	+	hydrogen
ethanol	+	sodium	→	sodium ethoxide	+	hydrogen

- Reaction with oxidising agents to form carboxylic acids. They will also form carboxylic acids by microbial action:

alcohol	+	oxidising agent	→	carboxylic acid	+	water
ethanol	+	oxidising agent	→	ethanoic acid	+	water

Vinegar is an aqueous solution and ethanoic acid is the main acid in vinegar.

Carboxylic Acids

Carboxylic acids are organic compounds containing carbon, hydrogen and oxygen. They contain the functional group –COOH. Carboxylic acids are members of a **homologous series**. The first three members of that series are shown in the table:

Name	Formula	Displayed Formula
Methanoic acid	HCOOH	H–C(=O)–O–H
Ethanoic acid	CH_3COOH	H–C(H)(H)–C(=O)–O–H
Propanoic acid	C_2H_5COOH	H–C(H)(H)–C(H)(H)–C(=O)–O–H

All carboxylic acids have names ending in -anoic acid.

Small-molecule carboxylic acids are soluble in water and dissolve to form acidic solutions (pH less than 7). Carboxylic acids are weak acids so they have a pH closer to 7 than a strong acid – e.g. hydrochloric acid – of the same concentration.

> HT All acids dissociate / ionise in water to produce H^+ ions, however because carboxylic acids only partially dissociate they are weak acids.
>
> Aqueous solutions of weak acids have a pH value closer to 7 than solutions of strong acids with the same concentration.

Reactions of Carboxylic Acids

Carboxylic acids react with carbonates to form a salt, carbon dioxide and water:

carboxylic acid	+	metal carbonate	→	salt	+	water	+	carbon dioxide
ethanoic acid	+	sodium carbonate	→	sodium ethanoate	+	water	+	carbon dioxide

Carboxylic acids react with alcohols in the presence of an acid catalyst (usually sulfuric acid) to produce an ester and water:

carboxylic acid	+	alcohol	⇌	ester	+	water
ethanoic acid	+	ethanol	⇌	ethyl ethanoate	+	water

Esters

Esters are organic compounds containing carbon, hydrogen and oxygen. They contain the functional group –COO–

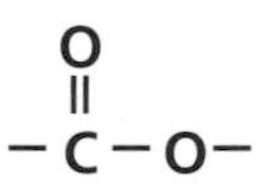

Esters are members of a **homologous series**. One of the members of that series is ethyl ethanoate:

Name	Formula	Displayed Formula
Ethyl ethanoate	$CH_3COOC_2H_5$	H–C(H)(H)–C(=O)–O–C(H)(H)–C(H)(H)–H

All esters have names ending in -anoate.

Esters are volatile compounds and because of their distinctive smells they are used in flavourings and perfumes.

Ethyl ethanoate is formed when ethanoic acid reacts with ethanol in the presence of an acid catalyst:

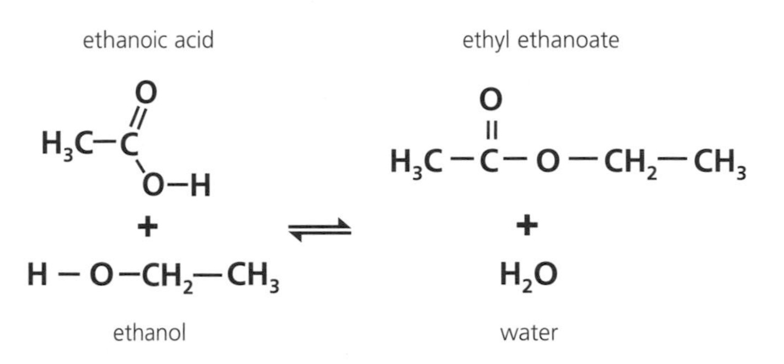

Uses of Alcohols, Carboxylic Acids and Esters

Compound	Use
Alcohols	Mainly used as a solvent and fuel (can be obtained from renewable sources, e.g. sugar cane or non-renewable sources (i.e. crude oil). Ethanol is the main alcohol in alcoholic drinks, e.g. beer and wine. It can be used as a starting material to make many other organic compounds, e.g. esters and carboxylic acids. Methanol is used as a raw material for many industrial processes. It is used in the manufacture of fuels, adhesives, foams, cosmetics and solvents.
Carboxylic acids	Ethanoic acid (vinegar) is used as a preservative and as a flavouring for food. Carboxylic acid functional groups are also present in citric acid (found in citrus fruit) and used as a preservative and in antacid powders. Long-chain carboxylic acids are known as fatty acids and are found in foods.
Esters	Used in perfumes and food flavourings for their sweet smell. Examples include flavouring in sweets, lavender oil and in perfumes. Occur widely in nature but many esters used are synthetic. Esters are also found in solvents and adhesives.

C3 Exam Practice Questions

1. Ammonia is produced in the Haber process.

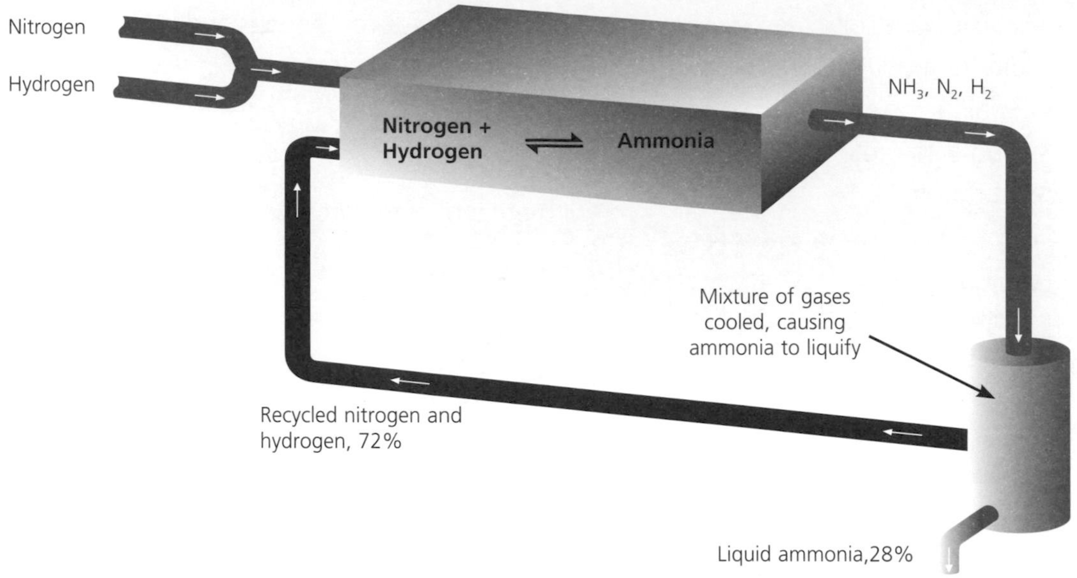

a) What are the sources of the two raw materials, nitrogen and hydrogen? **(2 marks)**

b) Write down a balanced symbol equation, including state symbols, for the reaction taking place. **(2 marks)**

c) At what temperature and pressure does the reaction take place? **(2 marks)**

d) What is the purpose of the iron catalyst? **(1 mark)**

e) Ammonia can be used in the production of nitrogen-based fertilisers, which can increase crop yields for farmers. Name one problem that using nitrogen-based fertilisers might cause. **(1 mark)**

2. Alcohols are organic compounds containing hydrogen, carbon and oxygen atoms.

a) What is the general formula for the alcohols? **(1 mark)**

b) What is the functional group of the alcohols? **(1 mark)**

c) Methanol can be burned as a fuel. Complete the balanced equation shown below: **(2 marks)**

$$CH_3OH + \ldots\ldots\ldots\ldots \longrightarrow CO_2 + \ldots\ldots\ldots\ldots$$

d) When an alcohol reacts with a carboxylic acid, another type of organic compound is produced. Give the name of this type of compound and one general use for them. **(2 marks)**

HT 3. A student carries out a titration using sulfuric acid (H_2SO_4) and sodium hydroxide (NaOH). $0.25dm^3$ of sodium hydroxide is neutralised by $0.75dm^3$ of sulfuric acid that has a concentration of $0.5mol/dm^3$.

a) Write the balanced symbol equation for this neutralisation reaction. **(2 marks)**

b) What is the concentration of the sodium hydroxide solution in this titration? **(2 marks)**

Unit 1

1. a) 2,8,7

b) It has 7 electrons in its outermost shell / energy level.

2. a) It is a mineral / mixture of minerals / rock from which it is economically viable to extract a metal.

b) Reduction; Less; Electrolysis

3. a) Hydrogen and carbon (only).

b) C_nH_{2n+2}

c) i) Heat; A catalyst

ii) Cracking

iii) (Short-chain) alkene

iv) Use bromine water; Bromine water will change from orange to colourless with the alkene but remain orange with the alkane (i.e. bromine is decolourised).

4. a) $4Na + O_2 \longrightarrow 2Na_2O$

b) $H_2SO_4 + 2NaOH \longrightarrow Na_2SO_4 + 2H_2O$

Unit 2

1. a)

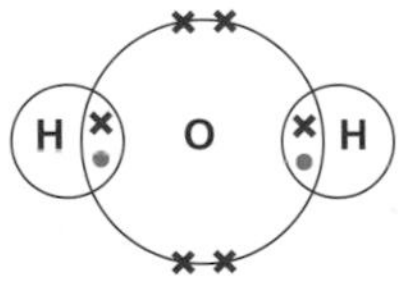

b) Covalent

2. When molten or when dissolved in water / in aqueous solution **[1 mark]** as ions are free to move. **[1 mark]**

3. a) Magnesium + hydrochloric acid $\longrightarrow$ magnesium chloride + hydrogen **[1 mark for reactants; 1 mark for products]**

b) Magnesium powder has a much larger surface area than the ribbon **[1 mark]**; Which means that there are more collisions between magnesium and hydrochloric acid when using the powder. **[1 mark]**

c) Use a catalyst; Increase the temperature of the reaction; Increase the concentration of the hydrochloric acid. **[Any two for 2 marks]**

4. Oxidation is loss of electrons; Reduction is gain of electrons.

5. a) To reduce the melting point of aluminium oxide.

b) Carbon

c) The carbon electrode reacts with the oxygen produced at the positive electrode (forms carbon dioxide).

6. a) $Al^{3+} + 3e^- \longrightarrow Al$ **[1 mark for ions; 1 mark for balancing]**

b) $2O^{2-} \longrightarrow O_2 + 4e^-$ **[1 mark for ions; 1 mark for balancing]**

Unit 3

1. a) Nitrogen from liquid air; Hydrogen from natural gas and steam.

b) $N_2(g) + 3H_2(g) \rightleftharpoons 2NH_3(g)$ **[1 mark for reactants; 1 mark for products]**

c) 450°C; 200 atmospheres.

d) Catalyst; Increases the rate of both reactions.

e) Contamination of drinking water or eutrophication.

2. a) $C_nH_{2n+1}OH$

b) –OH

c) $CH_3OH + 1\frac{1}{2}O_2 \longrightarrow CO_2 + 2H_2O$ **[1 mark for reactants; 1 mark for products]**

d) Ester **[1 mark]**; Used as a flavouring in food / used in perfumes. **[1 mark]**

3 a) $H_2SO_4 + 2NaOH \longrightarrow Na_2SO_4 + 2H_2O$ **[1 mark for formula; 1 mark for balancing]**

b) Moles of H_2SO_4 used = 0.75 × 0.5 = 0.375

Moles of NaOH neutralised = 0.375 × 2 = 0.75

Concentration of NaOH = $\frac{0.75}{0.25}$ = 3mol/dm^3

[1 mark for working; 1 mark for answer]

Glossary

Acid – a solution that has a pH value lower than 7

Activation energy – the minimum energy required for a reaction to occur

Alkali – a soluble base; pH value of solution higher than 7

Alkali metals – elements in Group 1 of the Periodic Table, including lithium, sodium and potassium

Alkane – a saturated hydrocarbon (C_nH_{2n+2})

Alcohol – organic compounds containing the functional group –OH ($C_nH_{2n+1}OH$)

Alkenes – an unsaturated hydrocarbon containing at least one carbon–carbon double bond (C_nH_{2n})

Alloy – a mixture of two or more metals or a mixture of one metal and a non-metal

Atom – the smallest part of an element that can enter into chemical reactions

Atomic number – the number of protons in an atom

Base – a compound that can neutralise an acid

Biofuel – a fuel produced from plant material

Bond energies – the energy required to break bonds or the energy given out when bonds are formed

Calorie – a unit of heat energy; the energy required to raise the temperature of 1g of water by 1°C

Calorimetry – the measurement of quantities of heat energy

Carboxylic acid – organic compounds containing the functional group –COOH

Catalyst – a substance that increases the rate of a chemical reaction while remaining chemically unchanged itself

Chemical formula – a way of showing the elements present in a substance

Chemical reaction – a process in which one or more substances are changed into others

Compound – a substance consisting of two or more different elements that are chemically combined

Covalent bond – a bond between two non-metal atoms in which both atoms share one or more pairs of electrons

Decompose – to break down

Distillation – the process of boiling a liquid and condensing its vapours

Electrodes – pieces of metal or carbon that allow electric current to enter and leave during electrolysis

Electrolysis – the process by which an electric current causes a solution to undergo chemical decomposition into its elements

Electron – a negatively charged particle orbiting the nucleus of an atom. In a neutral atom the number of electrons equals the number of protons.

Element – a substance that consists of only one type of atom

Emulsion – a liquid dispersed in another liquid, e.g. oil and water mixed together

Endothermic reaction – a reaction that takes in energy (usually heat) from the surroundings

Equilibrium – a chemical reaction in which the rate of the forward and backward reactions is the same

Ester – organic compound containing the functional group –COO–

Fossil fuel – a fuel formed in the ground over millions of years from the remains of dead plants and animals; a non-renewable fuel

Exothermic reaction – a reaction that gives out energy (usually heat) to the surroundings

Fuel – a substance that releases heat or energy when combining with oxygen

Group – a vertical column in the Periodic Table

Haber process – reaction in which ammonia (NH_3) is formed from nitrogen and hydrogen

Halogens – elements in Group 7 of the Periodic Table, including fluorine, chlorine, bromine and iodine

Hard water – water containing dissolved compounds (usually of calcium and magnesium) that prevent soap from lathering

Hydrocarbon – a compound containing only hydrogen and carbon

Hydrogenation – the process in which hydrogen is used to harden vegetable oils

Ion – a charged particle (negative or positive) formed when atoms gain or lose electrons

Ionic bond – the electrostatic attraction between oppositely charged ions; formed between metal atoms and non-metal atoms.

Isotopes – atoms of the same element that have the same number of protons (atomic number) but different numbers of neutrons (to give different mass numbers)

Joule – a unit of energy and work

Mass number – the number of protons plus the number of neutrons in an atom

Glossary

Metallic bond – the electrostatic attraction between delocalised electrons and positive metal ions; occurs between metal atoms only

Mixture – two or more different substances that are not chemically combined

Mole (mol) – one mole contains 6×10^{23} particles

Molecule – the simplest structural unit of an element or covalent compound

Monomer – the small molecules that combine to make up a polymer

Nanomaterials – materials with a very small grain size

Neutralisation – a reaction between an acid and a base that forms a neutral solution

Neutron – a particle in the nucleus of an atom; has no electrical charge

Noble gases – elements in Group 0 of the Periodic Table, including helium, neon and argon

Non-biodegradable – a substance that does not decompose naturally

Nucleus – the small central core of an atom, consisting of protons and neutrons

Ore – a naturally occurring mineral from which a metal can be extracted (if economically viable)

Oxidation – a reaction involving the gain of oxygen or loss of electrons

Period – a horizontal row in the Periodic Table

Polymer – a giant long-chain molecule made by combining lots of monomers

Polymerisation – the reaction in which a polymer is formed by joining together lots of monomers

Precipitation – when two solutions react to make an insoluble solid

Proton – a positively charged particle in the nucleus of an atom

Redox reaction – a reaction when an oxidation and reduction reaction occur at the same time

Reduction – a reaction involving the loss of oxygen or gain of electrons

Relative atomic mass (A_r) – the average mass (in atomic mass units) of the isotopes of an element; compares the mass of atoms of the element with the ^{12}C isotope

Relative formula mass (M_r) – the sum of the atomic masses of all the atoms in a compound

Reversible reaction – a reaction that goes both forwards and backwards, i.e. reactants form products and products form reactants

Smart material – a material that responds to changes in its environment

Soft water – water that reacts readily with soap to form a lather

Titration – chemical analysis used to determine the concentration of a known reactant

Transition elements – block of metallic elements in the middle of the Periodic Table, including copper and iron

Yield – the amount of a product obtained from a reaction

HT **Empirical formula –** the simplest whole number ratio of elements in a compound

Electronic Structure of the First Twenty Elements

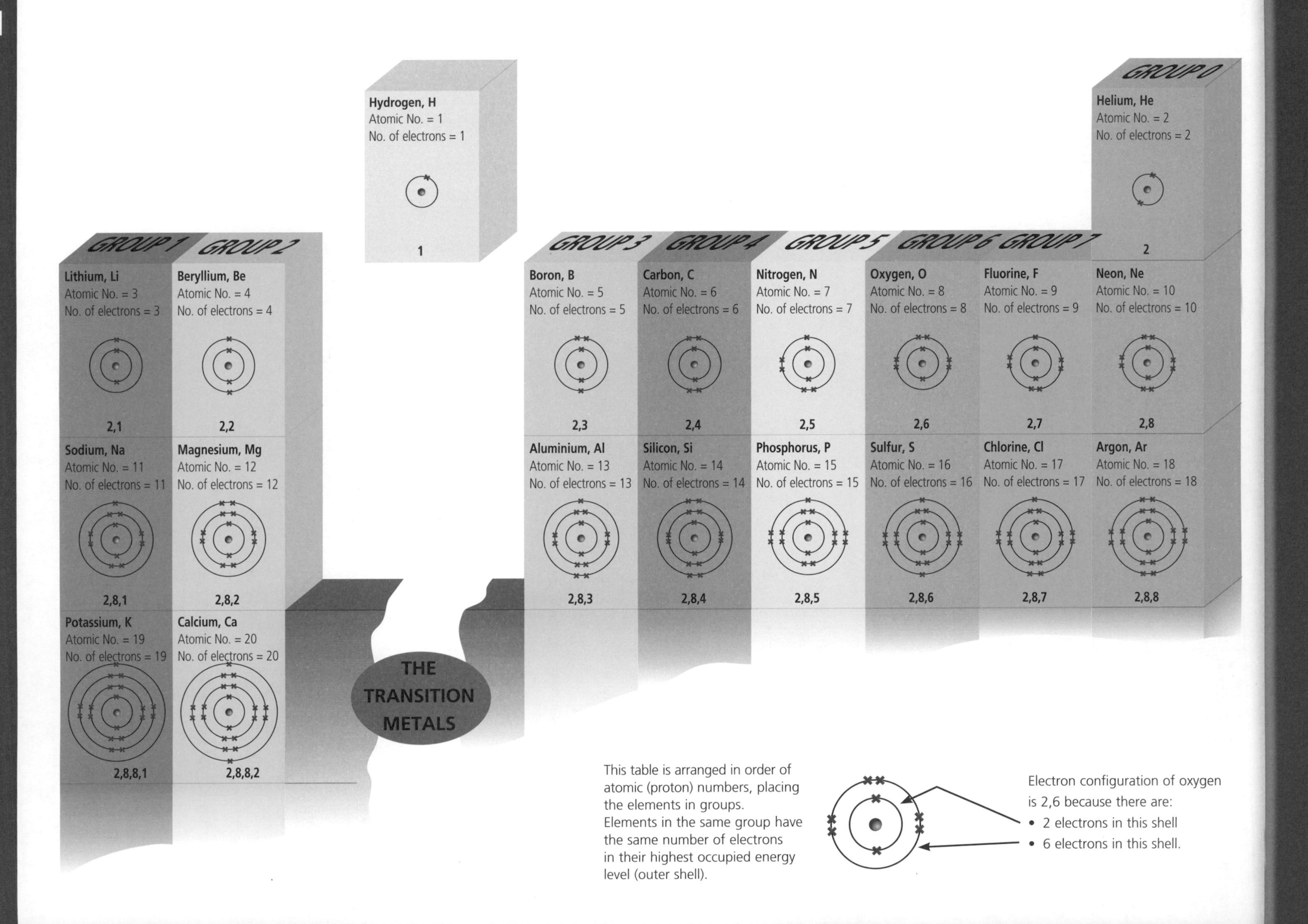

The Modern Periodic Table

Key

relative atomic mass **atomic symbol** name atomic (proton) number

1	2											3	4	5	6	7	0
							1 **H** hydrogen 1										4 **He** helium 2
7 **Li** lithium 3	9 **Be** beryllium 4											11 **B** boron 5	12 **C** carbon 6	14 **N** nitrogen 7	16 **O** oxygen 8	19 **F** fluorine 9	20 **Ne** neon 10
23 **Na** sodium 11	24 **Mg** magnesium 12											27 **Al** aluminium 13	28 **Si** silicon 14	31 **P** phosphorus 15	32 **S** sulfur 16	35.5 **Cl** chlorine 17	40 **Ar** argon 18
39 **K** potassium 19	40 **Ca** calcium 20	45 **Sc** scandium 21	48 **Ti** titanium 22	51 **V** vanadium 23	52 **Cr** chromium 24	55 **Mn** manganese 25	56 **Fe** iron 26	59 **Co** cobalt 27	59 **Ni** nickel 28	63.5 **Cu** copper 29	65 **Zn** zinc 30	70 **Ga** gallium 31	73 **Ge** germanium 32	75 **As** arsenic 33	79 **Se** selenium 34	80 **Br** bromine 35	84 **Kr** krypton 36
85 **Rb** rubidium 37	88 **Sr** strontium 38	89 **Y** yttrium 39	91 **Zr** zirconium 40	93 **Nb** niobium 41	96 **Mo** molybdenum 42	[98] **Tc** technetium 43	101 **Ru** ruthenium 44	103 **Rh** rhodium 45	106 **Pd** palladium 46	108 **Ag** silver 47	112 **Cd** cadmium 48	115 **In** indium 49	119 **Sn** tin 50	122 **Sb** antimony 51	128 **Te** tellurium 52	127 **I** iodine 53	131 **Xe** xenon 54
133 **Cs** caesium 55	137 **Ba** barium 56	139 **La*** lanthanum 57	178 **Hf** hafnium 72	181 **Ta** tantalum 73	184 **W** tungsten 74	186 **Re** rhenium 75	190 **Os** osmium 76	192 **Ir** iridium 77	195 **Pt** platinum 78	197 **Au** gold 79	201 **Hg** mercury 80	204 **Tl** thallium 81	207 **Pb** lead 82	209 **Bi** bismuth 83	[209] **Po** polonium 84	[210] **At** astatine 85	[222] **Rn** radon 86
[223] **Fr** francium 87	[226] **Ra** radium 88	[227] **Ac*** actinium 89	[261] **Rf** rutherfordium 104	[262] **Db** dubnium 105	[266] **Sg** seaborgium 106	[264] **Bh** bohrium 107	[277] **Hs** hassium 108	[268] **Mt** meitnerium 109	[271] **Ds** darmstadtium 110	[272] **Rg** roentgenium 111	Elements with atomic numbers 112–116 have been reported but not fully authenticated						

*The Lanthanides (atomic numbers 58–71) and the Actinides (atomic numbers 90–103) have been omitted.

Cu and **Cl** have not been rounded to the nearest whole number.

Index